Energy-Efficient VLSI Architectures for Real-Time
and 3D Video Processing

Michael Stefano Fritz Schaffner

Energy-Efficient VLSI Architectures for Real-Time and 3D Video Processing

Hartung-Gorre Verlag Konstanz
2018

Reprint of Diss. ETH No. 24261

SERIES IN MICROELECTRONICS   VOLUME 235

edited by   Qiuting Huang
            Andreas Schenk
            Mathieu Maurice Luisier
            Bernd Witzigmann

Bibliographic information published by Die Deutsche National-bibliothek

Die Deutsche Nationalbibliothek lists this publication in the Deutsche National-bibliografie; detailed bibliographic data is available on the Internet at: http://dnb.d-nb.de

First edition 2018

HARTUNG-GORRE VERLAG KONSTANZ
http://www.hartung-gorre.de

ISSN 0936-5362
ISBN-10: 3-86628-624-4
ISBN-13: 978-3-86628-624-5

Acknowledgements

First of all, I would like to thank my supervisor, Hubert Kaeslin, and my co-supervisors, Aljoscha Smolic and Luca Benini. I appreciate all the insightful discussions, your constructive criticism and effort to push me to the next level. The last four years have been intense, but I am grateful for everything I learned during this time. Hubert, thanks for sticking with me until the conclusion of my thesis. I wish you all the best for the time after your retirement!

Further, I thank Markus Gross from Disney Research for supporting the collaboration with Disney and thereby giving me the opportunity to work on an interesting and multifaceted project – and, at the same time, to see behind the scenes of the Disney magic.

I thank all my collaborators, Michael Gautschi, Frank Gürkaynak, Florian Scheidegger, Lukas Cavigelli, Pierre Greisen, Simon Heinzle, Pascal Hager and Youri Popoff for the fruitful cooperation that led to several joint contributions. In particular, I would like to thank Frank for his support and guidance during times where things were not going so smoothly. Also, I would like to thank Norbert Felber for motivating me to pursue ASIC design back in days of my graduate studies. Norbert, I guess I wouldn't be were I am now had I not visited that particular PPS course many years ago. Big thanks also go to all other office mates I have not mentioned so far. I will miss all the interesting and insightful discussions (especially the highly intellectual office trash talk) and wish you good luck on your future paths!

Last, I would like to thank all my friends and in particular my family, Benjamin, Heinz and Christine Schaffner for their unconditional support.

v

Abstract

Multiview autostereoscopic displays (MADs) make it possible to view video content in 3D without wearing special glasses, and such displays have recently become available. The main problem of MADs is that they require several (typically 8 or 9) views, while most of the 3D video content is in stereoscopic 3D today. To bridge this content-display gap, the research community started to devise automatic multiview synthesis (MVS) methods. Common MVS methods are based on depth-image-based rendering, where a dense depth map of the scene is used to reproject the image to new viewpoints. Although physically correct, this approach requires accurate depth maps and additional inpainting steps. Our work uses an alternative conversion concept based on image domain warping (IDW) which has been successfully applied to related problems such as aspect ratio retargeting for streaming video, and disparity remapping for depth adjustments in stereoscopic 3D content. IDW shows promising performance in this context as it only requires robust, sparse point-correspondences and no inpainting steps. However, MVS, using IDW as well as alternative approaches, is computationally demanding and requires real-time processing – yet such methods should be portable to end-user and even mobile devices to develop their full potential. To this end, this thesis investigates efficient algorithms and hardware architectures for a variety of subproblems arising in the MVS pipeline.

We begin by evaluating binary descriptors for identifying corresponding points in rectified stereo settings, followed by the design of a complete feature extraction and matching system. The system employs the SKB descriptor and distinguishes by a high matching accuracy and descriptor throughput. For other settings with more general viewing conditions typically arising in applications such as optical flow estimation, we design a new binary feature descriptor termed BOOM which significantly outperforms state-of-the-art descriptors. This descriptor is used to significantly accelerate a recent patch match method to efficiently calculate optical flow vectors.

IDW involves solving large, regularized least-squares problems to compute the warping functions. Hence, we discuss different types of sparse linear solvers with a focus on energy efficiency. A basic estimation framework has been developed to compare different solver architecture variants. We highlight potential tradeoffs to guide architectural and system level decisions, and show that in 28 nm technology, complete on-chip solvers without external DRAM are perfectly feasible even for large matrices, enabling high quality MVS in real-time.

MADs employ non-orthogonal sampling patterns to spatially multiplex different subviews which can create disturbing aliasing artifacts unless properly accounted for. We describe these effects and compare different resampling strategies with a real-time focus. Our assessment shows large differences in complexity for similar image quality, and provides insightful recommendations for selecting a suitable strategy.

Further, we investigate spatio-temporal, edge-aware (STEA) filtering methods that recently received increased attention due to their ability to efficiently approximate image-domain problems in a temporally consistent manner. The STEA pipeline presented accurately calculates optical flow, competing with related state-of-the-art methods. The method has the potential to improve the quality and efficiency of image-based processing algorithms such as IDW significantly, since STEA filtering scales better with the number of pixels in the image.

Finally, we investigate the use of logarithmic number systems (LNS) as an energy-efficient alternative to floating-point arithmetic for applications where fixed-point arithmetic does not suffice. These efforts lead to the development of a framework able to generate accurate and approximate LNS units. By integrating these units into a multi-processor system, we show that significant energy-efficiency gains can be achieved for specific signal and image processing kernels.

The experience collected in the domains of feature extraction, linear solvers and resampling techniques for MADs culminate in the realization of a hybrid FPGA/ASIC prototype for real-time MVS. With an estimated silicon area of $8.5\,\text{mm}^2$ and a power dissipation of ~$1.3\,\text{W}$ in 28 nm CMOS technology, the hardware accelerator enables portable and energy-efficient MVS. The demonstrator system serves as a proof of concept, paving the way towards completely integrated consumer electronic devices, e.g., as part of a system on chip with possible deployment in TV sets, set-top boxes or mobile devices.

Zusammenfassung

Autostereoskopische Multiview Displays (MADs), welche vor kurzem kommerziell erhältlich geworden sind, ermöglichen es 3D Video-Inhalte ohne spezielle Brille wahrzunehmen. Das Problem dabei ist, dass diese MADs mehrere (typischerweise 8 oder 9) Ansichten benötigen, während die meisten 3D Videoinhalte heutzutage nur in stereoskopischem Format verfügbar sind. Um diesem Mangel an bildschirmgerechten Videoinhalten entgegenzuwirken, hat man vor kurzem begonnen, Methoden für die automatische Multiview-Synthese (MVS) zu entwickeln. Gängige Methoden benutzen eine dichte Tiefenkarte der Szene um mittels tiefenbasiertem Rendering neue Ansichten zu generieren. Dieser Ansatz ist zwar physikalisch korrekt, erfordert jedoch genaue Tiefenkarten und zusätzliche Inpainting Schritte. Im Gegensatz dazu befassen wir uns in dieser Arbeit mit einem alternativen Konvertierungskonzept names Image Domain Warping (IDW), welches bereits erfolgreich auf verwandte Probleme angewendet wird. Dazu gehören z.B. die Anpassung des Seitenverhältnisses von Videosequenzen, sowie die Änderung von Disparitäten um Tiefenanpassungen in stereoskopischen 3D Inhalten vorzunehmen. IDW liefert in diesem Kontext vielversprechende Ergebnisse, da es nur robuste, dünnbesetzte Punktkorrespondenzen und keine Inpainting Schritte erfordert. Allerdings ist MVS mittels IDW rechnerisch anspruchsvoll und erfordert Echtzeit-Verarbeitung. Dies ist nicht leicht zu bewerkstelligen, da solche Methoden nahe beim Endbenutzer und auf mobilen Geräten eingesetzt werden müssen um ihr volles Potential zu entfalten. Aus diesen Gründen erforschen wir in dieser Arbeit effiziente Algorithmen und Hardware Architekturen für eine Vielzahl von Teilproblemen, die bei der MVS entstehen.

Als erstes untersuchen wir verschiedene binäre Deskriptoren für die Erkennung zusammengehöriger Punkte in rektifizierten Stereobildern. Dabei entwerfen wir ein komplettes Feature Erkennungs- und Zuordnungssystem, welches auf dem SKB-Deskriptor basiert. Dieses System zeichnet sich durch eine hohe Zuordnungsgenauigkeit sowie einen grossen Deskripor Durchsatz aus. Für andere Situationen mit allgemeineren Betrachtungsbedingungen, die z.b. bei der Schätzung des optischen Flusses auftreten, entwickeln wir einen neuen binären Deskriptor namens BOOM, der die Leistung moderner Deskriptoren deutlich übertrifft. Dieser Deskriptor wird verwendet, um eine kürzlich vorgeschlagene Patch-Match Methode für die Berechnung optischer Strömungsvektoren signifikant zu beschleunigen.

Um die Warping-Funktionen zu berechnen, müssen beim IDW grosse regularisierte Probleme der kleinsten Quadrate gelöst werden. Deshalb diskutieren wir verschiedene Arten von dünnbesetzten linearen Solvern mit einem Fokus auf Energieeffizienz und entwickeln ein einfaches Schätzungsmodell, um verschiedene Architekturvarianten zu vergleichen. Wir beleuchten verschiedene Aspekte um Entscheidungen auf der Architektur- und Systemebene zu lenken und zeigen dabei auf, dass in der 28 nm CMOS Technologie komplette on-chip Solver ohne externes DRAM auch für grosse Matrizen realisierbar sind – was ein Schlüssel zur hochwertigen MVS in Echtzeit darstellt.

Bei gängigen MADs werden verschiedene räumliche Ansichten mittels verschachtelten Abtastungsmustern gebündelt, was störende Aliasing Artefakte erzeugen kann falls das Bildmaterial nicht korrekt gefiltert wird. Wir beschreiben diese Effekte und vergleichen unterschiedliche Resampling-Strategien mit einem Fokus auf Echtzeitverarbeitung. Die Evaluation zeigt mögliche Kompromisse zwischen Komplexität und Bildqualität auf und liefert eine Empfehlung für die Auswahl einer geeigneten Resampling Methode.

Darüber hinaus untersuchen wir räumlich-zeitliche, kantensensitive (STEA) Filtermethoden, die neulich zur effizienten Approximation verschiedener Bilddomänenprobleme in einer zeitlich konsistenten Weise vorgeschlagen wurden. Die hier entwickelte STEA-Methode ist in der Lage, optischen Fluss genau zu berechnen und konkurriert mit verwandten modernen Methoden. Das Verfahren hat das Potential, die Qualität und Effizienz von bildbasierten Algorithmen wie IDW signifikant zu verbessern, da es besser mit der Anzahl der Pixel skaliert.

Schliesslich untersuchen wir die Verwendung von logarithmischen Zahlensystemen (LNS) als energieeffiziente Alternative zur Gleitkomma Arithmetik für Anwendungen, bei denen Fixpunkt Arithmetik nicht ausreichend ist. Dies führt zur Entwicklung eines Frameworks für die Erzeugung exakter und approximativer LNS-Rechenwerke. Mittels Integration dieser Einheiten in ein Multiprozessor-System zeigen wir, dass für spezifische Signal- und Bildverarbeitungsunterprogramme eine erhebliche Effizienzsteigerung erzielt werden kann.

Die gesammelten Erfahrungen aus den Bereichen der Feature Erkennung, linearen Solvern und Resampling Techniken für MADs führen zur Realisierung eines hybriden FPGA/ASIC-Prototyps der in der Lage ist, MVS in Echtzeit zu betreiben. Mit einer geschätzten Siliziumfläche von $8.5\,\text{mm}^2$ und einer Verlustleistung von ~$1.3\,\text{W}$ in $28\,\text{nm}$ CMOS-Technologie ermöglicht der Hardware-Beschleuniger energieeffiziente MVS bei kompakten Formfaktor. Das Demonstratorsystem belegt die Machbarkeit des Konzeptes und ebnet den Weg zu vollständig integrierten elektronischen Endgeräten, z.B. als Teil eines System-on-Chip mit möglichem Einsatz in Fernsehgeräten, Set-Top-Boxen oder mobilen Geräten.

Contents

Chapter 1

Introduction

1.1 Motivation

Over the last decade, video capture, post-processing and distribution technologies for stereoscopic 3D (S3D) content have become mature enough for broad commercialization [1–3]. Coupled with the box office success of S3D movies, this has brought on renewed interest in the development of S3D-capable consumer-electronic devices such as TV sets. However, most such devices require the viewers to wear some sort of shutter- or polarization glasses, which is often regarded as an inconvenience [4]. Recently, so-called multiview autostereoscopic displays (MADs) [5, 6] have become commercially available[1]. These are able to project several views of a scene simultaneously – enabling a glasses-free 3D experience and a limited motion parallax effect in horizontal direction. However, appropriate content for such displays is virtually inexistent since storage and transmission of HD content with more than two views is impractical and even infeasible in some cases. The fact that each MAD has different parameters exacerbates the problem. In order to bridge this *content-display gap*, multiview synthesis (MVS) methods [3, 7–9] have been devised over the past couple of years. These methods are able to generate M virtual views from a small set of N input views, as shown in Figure 1.1.

---

1. An review of 3D display technologies and their history can be found in [10–12].

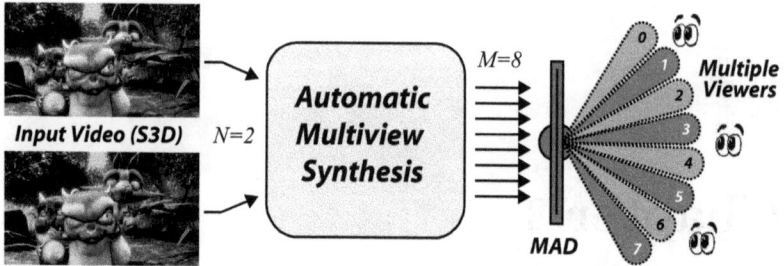

FIGURE 1.1: MVS generates a large number of synthetic views (e.g., $M = 8, 9$ for current MADs) from a small number of input views ($N = 2$ for S3D). A MAD is a display with a *lenticular sheet* mounted on top. Micro lenses project columns of subpixels in slightly different directions in such a way that the rays converge at several viewing positions. Images copyright © 2008, BLENDER FOUNDATION, www.bigbuckbunny.org.

Common MVS methods are based on depth image based rendering (DIBR) [3, 13], where a dense depth map of the scene is used to reproject the image to new viewpoints. Although physically correct, this approach requires accurate depth maps and additional inpainting[2] steps. Our work uses an alternative conversion concept suggested by [14] which is based on image-domain warping (IDW) [15]. The IDW framework allows to locally distort image regions via a non-linear, two-dimensional transformation obtained by solving a least-squares (LS) problem. The constraints for this problem are formulated using image features extracted from the input images. This technique is convenient as it does not rely on pixel-dense depth, but only on robust, sparse correspondences. Further, no inpainting is required which is still an algorithmically difficult step of DIBR-based MVS [16].

Warping-based MVS, as well as alternative approaches, are computationally intensive – yet they should run efficiently in real-time and should be portable to end-user devices to develop their full potential. However, high-performance GPU-based systems with enough throughput to guarantee real-time performance are typically not suited for these applications due to strict constraints on cost, size and energy efficiency. From a practical point of view, it is therefore essential to investigate hardware-friendly algorithms and approximations, as well

---

2. In disparity remapping and MVS applications, changes in camera perspective can lead to *disocclusions* where no information about the depicted scene is available. Such 'holes' can be filled in using *inpainting techniques*, that complete missing parts using extrapolation or statistical models [17, 18].

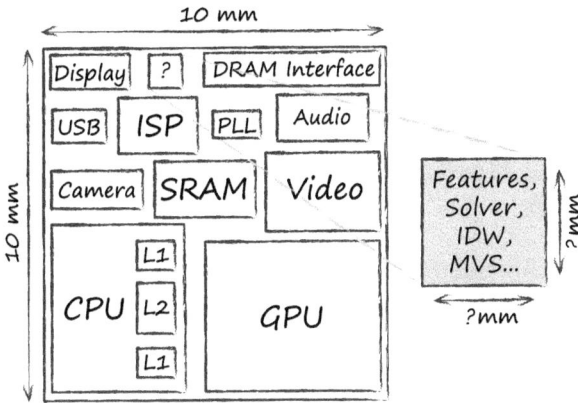

FIGURE 1.2: A typical system on chip (SoC) as it can be found in many consumer electronic devices today [19–22]. Heterogeneity and specialization are a key enabler for energy-efficient operation in TV-sets, set-top boxes, mobile phones and the like.

as dedicated hardware architectures in order to successfully deploy these technologies.

The investigation and development of specialized hardware IP for real-time, 3D video processing is also relevant in light of the maturing capability to integrate specialized accelerators together with general purpose processors on the same chip. According to GARTNER, INC. [23], such application-specific systems on chip (SoCs) [24, 25] represent '*the most important type of semiconductor device since the development of the microprocessor*', and have become the enablers of todays consumer electronic devices such as smart phones, tablets, set-top boxes and multimedia players[3]. As illustrated in Figure 1.2, a key aspect of SoCs is the heterogeneous integration of a variety of specialized components ranging from peripheral blocks for USB and camera management, to video CODECs like AVC and HEVC, and application specific DSPs like vision processing units (VPUs) [26] and image signal processors (ISPs). However, the algorithms involved require a detailed analysis in order to understand the main tradeoffs, and to come up with suitable and efficient hardware architectures.

---

3. Prominent examples for SoCs geared towards mobile devices are APPLE's A series, QUALCOMM's Snapdragon [21], SAMSUNG's Exynos series [19] and NVIDIA's

To this end, we investigate efficient algorithms and hardware architectures for a variety of subproblems in the context of IDW-based MVS, including:

- Efficient accelerators and novel binary descriptors for the extraction of sparse point-correspondences used for depth and optical flow estimation.
- Linear solver architectures for solvingLS problems emanating from IDW applications in real-time.
- Fast spatio-temporal edge-aware (STEA) filtering algorithms for approximate solution of regularization problems and temporally stable feature maps.
- Different IDW resampling strategies that include display anti-aliasing for MADs.
- Energy-efficient high dynamic range (HDR) arithmetic for accelerating non-linear functions present in several signal processing and computer vision problems.

As part of these investigations, we design and implement a hybrid ASIC/FPGA-based demonstrator system able to perform automatic MVS in real-time. The hardware IPs developed are used to estimate circuit complexity and energy efficiency in modern CMOS nodes. The system serves as a proof-of-concept and paves the way towards completely integrated solutions, e.g., as part of a future SoC with possible deployment in TV sets, set-top boxes or mobile devices.

The subsequent section gives a brief introduction to the IDW pipeline, followed by a thesis overview in Section 1.3. A summary of prior work and the main contributions of the present PhD thesis are given in Sections 1.4 and 1.5, respectively. Section 1.6 finally provides a list of publications by the author.

1.2   Image Domain Warping

IDW is an image-based technique for non-linearly transforming two-dimensional images and texture data. The transformation is typically

---

Tegra series [20]. Another interesting development are so-called platform FPGAs containing application processors alongside a programmable FPGA matrix [27].

available in the form of a discretized grid[4] as shown in Figure 1.3b). In the early days of digital image and video processing, IDW emerged as one of the first prominent techniques to create visual special effects [15]. Later, it has found widespread use in image editing software such as ADOBE Photoshop for image deformation and adaption. However, automatic generation of warping functions is a relatively new concept and has been introduced to address several related *display-content gap* problems, where streaming video content is adapted for view on the end-user device. Important applications are:

1. *Automatic video retargeting* [29–31] deals with changing the aspect ratio of video streams (e.g., from 4:3 to 16:9), where the content is transformed in content-adaptive manner such as to minimize distortions in important image regions.

2. *Non-linear disparity remapping* [32] allows to manipulate the depth distribution of a scene with the goal of keeping it within the display system's depth range, and to adapt 3D content to make it more comfortable for the user to view (e.g., by reducing extreme and uncomfortable depth values).

3. *Automatic MVS* [14], where a small set of input views (e.g. S3D) is converted into many virtual views appropriate for driving multiview (MV) devices such as MADs.

This thesis focuses on the third application. For more details on video retargeting, we refer to [33]. The second application is closely related to MVS since IDW is employed in a conceptually similar way to manipulate the depth of a scene. While it is not the focus of this thesis, most methods that are required for MVS (such as sparse point correspondence features and linear solvers) can be directly applied to disparity remapping too.

Figure 1.3 shows the IDW concept in the context of disparity remapping and MVS applications. An image warp can be pictured as a regular grid as shown in a) that is deformed by enforcing several constraints on it. The deformed grid, or *image warp*, shown in b) is then obtained by solving an LS problem in order to balance the different constraints. This warp can then be used to resample an input

---

4. Analytic representations, e.g. affine or projective transforms [28], can sometimes be used as well. Yet, in general, they do not provide enough degrees of freedom to represent complex transformations and hence grids are typically employed.

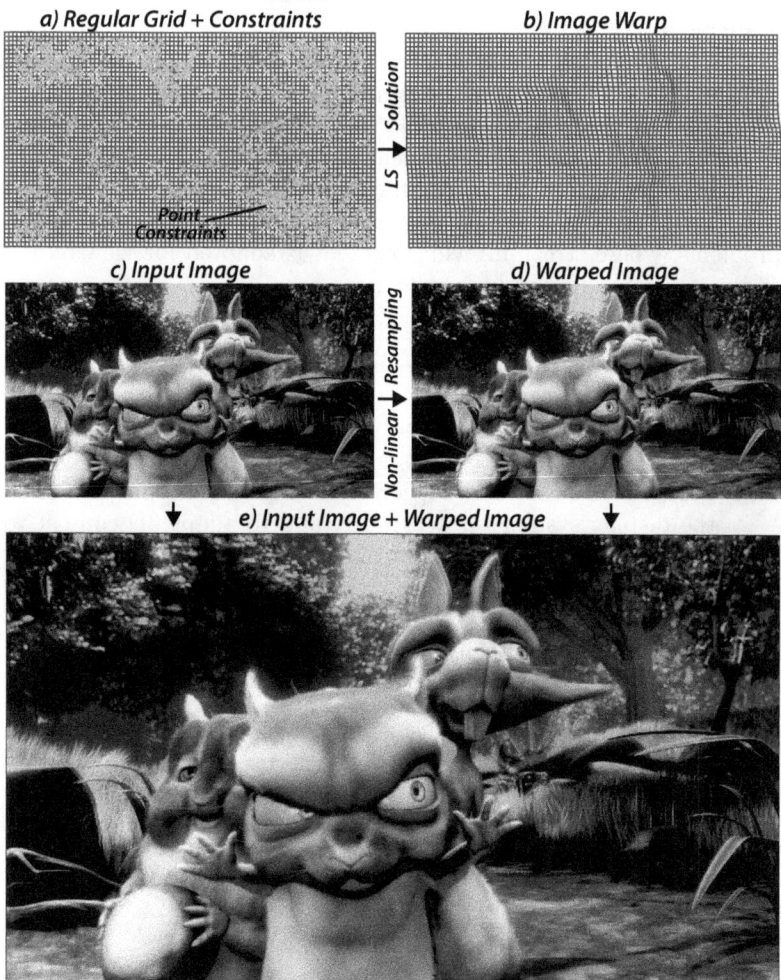

FIGURE 1.3: IDW concept in the context of disparity remapping and MVS applications: A regular grid a) is deformed by enforcing several constraints on it (in this example, point constraints are shown that enforce a certain position of the warp vertices). The *image warp* shown in b) is then obtained via solving an LS problem. This warp can be used to resample the input image c) in order to produce the warped image d). In this example, we slightly change the viewpoint of an image in order to produce the S3D pair shown in d) that is best viewed using anaglyph (red/blue) glasses.

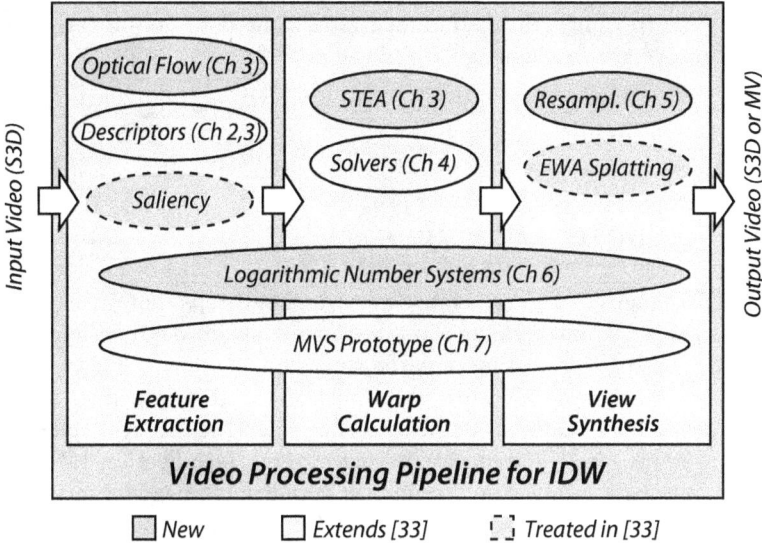

FIGURE 1.4: Thematic overview of the thesis. The topics are arranged in the context of a typical IDW flow for S3D video processing, and the colours indicate whether the contributions overlap with prior work [33], or whether they represent genuinely novel and independent contributions.

image c) in order to produce a warped image d). In this example, we slightly change the viewpoint of the original image in order to produce the S3D pair shown in d). The IDW concept has the big advantage that no disocclusion holes are created, and is leveraged in MVS to synthesize virtual views.

A typical IDW processing pipeline consists of three main stages, as illustrated in Figure 1.4:

1. The input footage is analyzed in order to reveal features such lines, sparse corresponding point pairs and visual saliency [34], which is an indicator for important information in the image.

2. These point correspondences are used to derive constraints on the warping grid (e.g. geometric relations are used to calculate the position of the sparse points after a slight perspective shift). These constraints are then used to set up and solve an LS problem where the variables are the unknown pixel positions.

3. The warping grid so obtained is then used to non-linearly resample the input image.

A more detailed algorithmic description of MVS using IDW can be
found in Sections 7.2 and 7.3.

1.3 Thesis Overview

All three stages in Figure 1.4 involve several substeps that are challenging to solve in real-time and in a resource-efficient manner, and therefore represent interesting research topics on their own. This section
gives an overview of our work performed on the IDW pipeline. While
prior work by P. Greisen [33] focused on general feasibility studies of
HW systems for IDW-based video retargeting (see Section 1.4), our
contributions focus on MVS, temporal stability of image features and
energy-efficient computation. The thesis consists of six main chapters,
summarized below. Figure 1.4 also provides a thematic arrangement
of the topics discussed.

Chapter 2 begins by evaluating binary descriptors for calculating
sparse point-correspondences, followed by the design of a complete
feature extraction and matching system. These features provide an
accurate and robust means for sparse depth-estimation, a crucial ingredient for IDW-based S3D processing and MVS.

In Chapter 3, we investigate so called STEA filtering methods that
recently received increased attention due to their ability to efficiently
solve or approximate important image-domain problems (such as optical flow estimation, sparse data upsampling, visual saliency computation and disparity estimation) in a temporally consistent manner.

Solving for the image warping function is a crucial step in IDW,
and requires a sparse linear system solver for LS problems. We discuss different solver types in Chapter 4, and present an architectural
exploration with focus on energy-efficiency.

Common MADs employ non-orthogonal sampling patterns in order to spatially multiplex all different subviews into one image. These
sampling arrangements can create disturbing aliasing artifacts unless
properly accounted for. Chapter 5 describes these aliasing effects and
compares different resampling strategies.

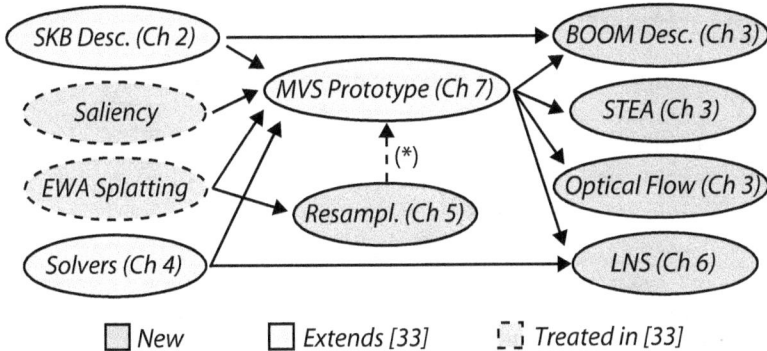

FIGURE 1.5: Chronological overview of the thesis that illustrates which topics influenced each other. The colours indicate whether the contributions overlap with prior work [33], or whether they represent genuinely novel and independent contributions. (*) Chapter 5 only partially influences the MVS prototype Chapter 7, as it also describes a novel technique that has not yet been integrated into the MVS demonstrator.

In Chapter 6, we consider the logarithmic number system (LNS) as an energy-efficient replacement for standard floating point (FP) arithmetic typically used in processing steps that require a higher dynamic range than what fixed-point arithmetic provides.

Using the knowledge collected from other chapters and prior work, we build the first IDW-based demonstrator system able to perform automatic MVS in real-time. Chapter 7 describes this system, provides a discussion of the visual quality of the results obtained, and compares the system with various related hardware components. The implemented IPs are also used to estimate the complexity of a completely integrated solution, e.g., as part of an SoC.

While all chapters are related to IDW, not all of the investigations have been carried out before building the MVS prototype. The chronological order of these chapters is illustrated in Figure 1.5, where arrows reflect dependencies among the chapters. Work on sparse features (semantic kernels binarized (SKB)), saliency, solvers and resampling methods (such as elliptical weighted average (EWA) splatting) culminated in the development of the MVS demonstrator. In turn, these efforts inspired a set of novel methods and techniques mainly treated in the STEA and LNS chapters. Due to the diversity of these topics, corresponding related work is discussed in detail at the beginning of each chapter.

1.4   Prior Work

To the authors knowledge, there does not exist much prior work on HW systems for IDW, except for the PhD thesis by P. Greisen [33] focusing on video retargeting. The present thesis is a continuation of P. Greisen's work within the DISNEY RESEARCH collaboration, and the relevant topics that overlap are briefly discussed below:

- Visual saliency is an important feature for all IDW applications, as it is used to guide the warp calculation. The efficient hardware architecture by P. Greisen [33] for an FFT-based algorithm [35] has been employed in this work.

- EWA splatting is a resampling algorithm well suited for IDW. Several hardware architectures optimized for single-view rendering have been jointly investigated by [33] and the author. Based on this prior work, an ASIC optimized for multiview rendering has been implemented by the author in the scope of his master thesis [36]. Testing and integration of this ASIC into the MVS prototype has been carried out later in preparation of the present PhD thesis. The ASIC architecture is briefly summarized in Chapter 7, but does not represent a main contribution.

- The feature extraction system (Chapter 2) contains a core part for interest point detection and descriptor calculation. This part is a joint contribution and has been briefly described in [33]. However, the descriptor performance evaluations and the complete extraction and matching system are new.

- An iterative and a direct linear solver for IDW problems have been evaluated and implemented on an FPGA in [33]. We extend this evaluation in Chapter 4 by looking into scalability aspects of different variants of iterative and direct solvers, and provide a design space exploration in 28 nm CMOS technology that also includes energy-efficiency estimates. The direct solver employed in the MVS prototype (Chapter 7) is different from the one implemented in [33] and contains several architectural improvements resulting in lower latency and higher throughput.

- While an architecture for a one-view synthesis system has been outlined in [33], no complete MVS system has been investigated and realized. The system presented in Chapter 7 can therefore be seen as a continuation of the early conceptual work in [33].

Another thesis by A. Chapiro [37] carried out in collaboration with DISNEY RESEARCH is concerned with perceptual enhancements for content to be viewed on novel displays such as MADs. For example, techniques have been developed that visually enhance the 3D experience provided by MADs, or that enable better utilization of the depth range provided by MADs by solving non-linear disparity remapping problems. While A. Chapiro's work is related to the present thesis, there is no overlap in terms of contributions.

The work on LNS[5] has been carried out jointly with M. Gautschi and published in [38–40]. While the author has concentrated on the evaluation and design of the arithmetic units implementing LNS, M. Gautschi focused on integration aspects in the multi-core processing system. A more detailed attribution of the contributions will be given in Chapter 6.

1.5 Contributions

The key contributions of this thesis are summarized below. More details can be found at the beginning of each corresponding chapter.

Evaluation and Development of Binary Feature Descriptors. We evaluate common binary descriptors in the context of rectified stereo settings, and show that the recently proposed binary descriptor termed SKB is an excellent choice in terms of accuracy. However, we also show that its performance drops in settings with larger differences in viewing positions, typically arising in applications such as optical flow-estimation. To this end, we design a new binary feature descriptor termed binarized octal orientation maps (BOOM) which significantly outperforms other state-of-the-art descriptors in more general viewing conditions.

Evaluation of Practical Resampling Strategies for MADs. We provide an evaluation of different resampling strategies for MADs.

---

5. While all other work in the present thesis has been carried out within the DISNEY RESEARCH collaboration, the work on LNS originates from the 'IcySoC' project, evaluated by the Swiss NSF and funded by Nano-Tera.ch with Swiss Confederation financing (http://www.nano-tera.ch/projects/451).

In particular, we compare the computational efficiency, memory requirements, and image quality of different resampling methods with a focus on real-time architectures. Our assessment shows large differences in complexity for similar image quality, and aims at providing a recommendation for selecting a suitable resampling strategy. The evaluation primarily targets the IDW framework, but the results can also be applied to DIBR.

An Efficient STEA Pipeline. We develop an efficient STEA filtering pipeline based on a recently proposed high-quality edge-aware (EA) filter, that exhibits desirable diffusion and halo-reduction characteristics. Using the mathematical properties of this filter, we reformulate its temporal extension as a causal, non-linear infinite impulse response (IIR) filter, which can be computed efficiently due to its incremental nature. We use the filter to bootstrap accurate optical flow required for the temporal extension. This is achieved by EA interpolation of a quasi-dense field of point-correspondences obtained with an improved *PatchMatch* algorithm, that employs a our new BOOM descriptor to find correspondences among subsequent frames. Our method is able to create temporally consistent results for a variety of applications such as optical flow estimation, sparse data upsampling, visual saliency computation and disparity estimation. We benchmark our optical flow method on the MPI Sintel dataset, where we currently achieve a Pareto optimal quality-efficiency tradeoff. Our method has significantly lower computational complexity, memory bandwidth and capacity requirements than previous STEA algorithms, thereby paving the way towards real-time software and hardware implementations in resource constrained settings.

A Design Space Exploration of Solver Architectures for IDW. The warp calculation step of the IDW pipeline involves the solution of a regularized LS problem using a linear solver. While prior work [33] focused on the evaluation and FPGA implementation of two discrete instances of a direct and an iterative solver, we consider the scalability and energy efficiency aspect of such methods in 28 nm CMOS technology. A basic estimation framework has been developed to compare three different architecture variants in terms of energy consumption, circuit area, throughput and precision. We highlight potential tradeoffs to guide architectural and system level decisions, and show that in 28 nm technology, complete on-chip solvers without external DRAM are perfectly feasible for matrix dimensions up to 128 k, enabling high quality IDW-based view synthesis in real-time.

A Generator and a Design Space Exploration for Logarithmic Number Units (LNUs). We develop a methodology to generate accurate and approximate LNUs capable of natively evaluating LNS additions and subtractions, typecasts (integer to float, float to integer), and special functions such as base-2 logarithms and exponentials, and trigonometric functions such as sine, cosine, tangent and their inverse. We provide a design space exploration of these units in the accuracy range between half- and single precision. Finally, we integrate differently parametrized LNUs as a shared unit into a multi-core RISC cluster and show comprehensive results of benchmark applications from different signal processing domains. These results demonstrate, that shared LNS can be an energy-efficient alternative to standard FP for the evaluation of complex non-linear functions.

A HW System for Automatic MVS in Real-time. We present the first IDW-based MVS hardware system able to convert S3D content into MV content appropriate for MADs in real-time, allowing to effectively bridge the content-display gap for this application. The system contains represents the culmination of this and prior efforts. Apart from addressing system integration aspects, the following important subsystems have been evaluated and developed in this thesis:

- A sparse feature extraction and matching subsystem using SKB descriptors. The system is distinguished by a high matching accuracy and throughput of up to 25 k points per frame from 720p stereo input at 42 fps. This ensures accurate point correspondences and high coverage of the scene – which is an important property for S3D and MVS applications.
- A micro code programmable ASIP for on-the-fly assembly of the IDW system matrices. This allows for more convenient specification of the LS matrices that encode the warping constraints.

- An efficient HW architecture for a sparse linear solver based on the Cholesky decomposition, that employs custom fused FP operators. The arithmetic and microarchitectural optimizations lead to low pipeline latency and high utilization of the datapath, therefore resulting in high throughput in terms of solve/s.

1.6 List of Publications

Most of the material covered in this thesis has been published within the DISNEY RESEARCH collaboration in the following conference and journal papers:

[41] M. Schaffner, P. Greisen, S. Heinzle, A. Smolic. Efficient Image Resampling for Multiview Displays. *Proc. of the International Conference on Acoustics, Speech and Signal Processing (ICASSP)*, Vancouver, CAN, 2013.

[42] M. Schaffner, P. Hager, L. Cavigelli, P. Greisen, F. K. Gürkaynak and H. Kaeslin. A Real-Time 720p Feature Extraction Core Based on Semantic Kernels Binarized. *Proc. of the International Conference on Very Large Scale Integration (VLSI-SoC)*, Istanbul, Turkey, 2013.

[43] M. Schaffner, F. K. Gürkaynak, A. Smolic, H. Kaeslin and L. Benini. An Approximate Computing Technique for Reducing the Complexity of a Direct-Solver for Sparse Linear Systems in Real-Time Video Processing. *Proc. of the Design Automation Conference (DAC)*, San Francisco, USA, 2014.

[44] M. Schaffner, P. A. Hager, L. Cavigelli, Z. Fang, F. K. Gürkaynak, P. Greisen, A. Smolic, H. Kaeslin and L. Benini. A Complete Real-Time Feature Extraction and Matching System Based on Semantic Kernels Binarized. *VLSI-SoC: At the Crossroads of Emerging Trends, IFIP Advances in Information and Communication Technology.* Springer Berlin Heidelberg, 2015.

[45] M. Schaffner, F. K. Gürkaynak, A. Smolic and L. Benini. DRAM or no-DRAM? Exploring Linear Solver Architectures for Image Domain Warping in 28 nm CMOS. *Proc. of the Design, Automation and Test in Europe Conference (DATE)*, Grenoble, France, 2015.

[46] M. Schaffner, F. K. Gürkaynak, H. Kaeslin, L. Benini and A. Smolic. Automatic Multiview Synthesis - Towards a Mobile System on a Chip. *Proc. of the Visual Communications and Image Processing Conference (VCIP)*, Singapore, 2015.

[47] M. Schaffner, F. K. Gürkaynak, H. Kaeslin, L. Benini and A. Smolic. Automatic Multiview Synthesis - Prototype Demo. *Proc. of the Visual Communications and Image Processing Conference (VCIP)*, Singapore, 2015.

[48] M. Schaffner, F. K. Gürkaynak, P. Greisen, H. Kaeslin, L. Benini and A. Smolic. Hybrid ASIC/FPGA System for Fully Automatic Stereo-to-Multiview Conversion using IDW. *Trans. on Circuits and Systems for Video Technology (TCSVT)*, 26(11) : 2093-2108, 2016.

[49] M. Schaffner, F. Scheidegger, L. Cavigelli, H. Kaeslin, L. Benini and A. Smolic. Towards Edge-Aware Spatio-Temporal Filtering in Real-Time. *Trans. on Image Processing (TIP)*, submitted, 2017.

The following publications covered by the present thesis are not affiliated with DISNEY RESEARCH:

[38] M. Gautschi, M. Schaffner, F. K. Gürkaynak and L. Benini. A 65 nm CMOS 6.4-to-29.2pJ/FLOP@0.8V Shared Logarithmic Floating Point Unit for Acceleration of Nonlinear Function Kernels in a Tightly Coupled Processor Cluster. *Proc. of the International Solid State Circuits Conference (ISSCC)*, San Francisco, USA, 2016.

[40] M. Schaffner, M. Gautschi, F. K. Gürkaynak and L. Benini. Accuracy and Performance Trade-offs of Logarithmic Number Units in Multi-Core Clusters. *Symposium on Computer Arithmetic (ARITH)*, Santa Clara, USA, 2016.

[39] M. Gautschi, M. Schaffner, F. K. Gürkaynak and L. Benini. An Extended Shared Logarithmic Unit for Non-linear Function Kernel Acceleration in a 65-nm CMOS Multi-Core Cluster. *Journal on Solid State Circuits (JSSC)*, vol. 52, no. 1, pp. 98-112, Jan 2017.

Prior work on IDW and EWA splatting by P. Greisen and the author (within the DISNEY RESEARCH collaboration):

[50] P. Greisen, R. Emler, M. Schaffner, S. Heinzle, and F.K. Gürkaynak. A General-Transformation EWA View Rendering Engine for 1080p Video in 130 nm CMOS. *Proc. of the International Conference on Very Large Scale Integration (VLSI-SoC)*, Santa Cruz, USA, 2012.

[51] P. Greisen, M. Schaffner, S. Heinzle, M. Runo, A. Smolic, A. Burg, H. Kaeslin, and M. Gross. Analysis and VLSI Implementation of EWA Rendering for Real-Time HD Video Applications. *Trans. on Circuits and Systems for Video Technology (TCSVT)*, 22(11) : 1577-1589, 2012.

[52] P. Greisen, M. Schaffner, D. Luu, V. Mikos, S. Heinzle, F.K. Gürkaynak, and A. Smolic. Spatially Varying Image Warping: Evaluations and VLSI Implementations. *VLSI-SoC: From Algorithms to Circuits and System-on-Chip Design.*, Springer Berlin Heidelberg, vol 418, pp. 64-87, 2013.

[36] M. Schaffner, P. Greisen, S. Heinzle, F.K. Gürkaynak, H. Kaeslin, and A. Smolic. MADmax: A 1080p Stereo-to-Multiview Rendering ASIC in 65 nm CMOS based on Image Domain Warping. *Proc. of the European Soli-State Circuits Conference (ESSCIRC)*, Bucharest, Romania, 2013.

Further publications by the author not explicitly covered in this thesis (not affiliated with DISNEY RESEARCH) are [53] and [54].

Chapter 2

Sparse Feature Extraction and Matching

2.1 Introduction

Current image and video processing pipelines commonly rely on image features in order to calculate sparse point correspondences among several images or frames. Over the past decade, numerous different algorithms have been devised. Earlier methods such as SIFT [55] and SURF [56] are costly to compute and the calculated descriptors consist of floating point entries that require a significant amount of memory – which renders them ill-suited for embedded devices or hardware implementations. This led to the development of more efficient *binary* descriptors such as BRIEF [57], BRISK [58], FREAK [59] and SKB [60]. These are less expensive to compute (e.g., BRIEF and SKB directly compute the binary descriptor pattern by means of intensity comparisons or thresholding), require less storage, and can be matched very efficiently using the Hamming distance.

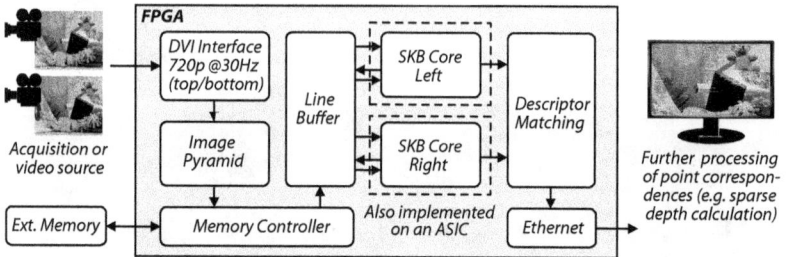

FIGURE 2.1: Overview of the stereo video feature detection system with two SKB cores and one matching unit. The 'Teddy' image shown here is from the dataset by [61].

In this chapter, we consider the calculation of a sparse disparity map from stereo video[1], as this is a crucial ingredient for several video processing methods such as non-linear depth adjustments [32] and automatic MVS [14]. We analyze the SKB descriptor [60], and show that it provides competitive results in restricted stereo vision settings. In addition, SKB is ideally suited for HW implementations since it is a binary descriptor that has low memory footprint and that only requires simple integer arithmetic. We develop an efficient hardware architecture for SKB feature extraction and matching for 720p stereo video (a system overview is shown in Figure 2.1). Due to the requirements of our target applications (MVS and stereo remapping), the feature extraction and matching parts have been designed to extract and match a much higher number of interest points per frame (up to 25 k) than what typical state-of-the-art architectures currently achieve (typically 512 - 2 k [62–64]). Instead of using a two-dimensional integral image[2]to compute the filter responses in the interest point detection stage, we develop an evaluation strategy which uses a local, one-dimensional integral image in order to overcome the large memory entries and the associated bandwidth that a two-dimensional integral image entails. The feature extraction core has been implemented and fabricated in 180 nm CMOS technology, and the whole system has been implemented on an ALTERA Stratix IV based FPGA evaluation board where it runs with up to 142 MHz, delivering a throughput of 42 fps (stereo video).

---

1. Optical flow estimation is a closely related problem that exhibits fewer geometrical constraints. Sparse features for flow estimation will be discussed in Chapter 3.

The system overview and our version of the SKB algorithm are explained in Section 2.3. A performance evaluation of the implemented configuration is shown at the end of that section. The hardware architecture is explained in Section 2.4.1. Implementation results and comparisons with related implementations are given in Section 2.5, while Section 2.6 draws the conclusions.

2.2 Related work

There exist many FPGA and ASIC implementations of SIFT [55] and SURF [56], such as [66–71]. FPGA or ASIC implementations of complete feature extraction and matching systems based on binary descriptors have only been published by Park et al. [62] and Wang et al. [63] so far. Park et al. [62] implemented a variant of BRIEF with FAST keypoints [72] on an ASIC that has a throughput of 94 full-HD frames per second with 512 extracted feature points per frame. Wang et al. [63] propose an FPGA system which is also based on BRIEF, but with SIFT keypoints. In their implementation, they achieve a feature throughput of 60 fps for 720p video with 2 k detected points. A relatively recent FPGA system based on non-binary SIFT descriptors also including a random sample consensus (RANSAC) filtering stage for outlier removal has been published in [64]. However, this system uses significantly more resources for similar throughput as their binary counterparts.

All systems mentioned are single view systems matching the extracted descriptors to the ones extracted from the previous frame. In contrast, our system operates on the left and right views simultaneously, and the constrained stereo camera set-up is exploited in order to match the descriptors from both views on the fly. Further, our system is capable of extracting and matching a much higher number of features per frame (15 k- 25 k), and is therefore tailored to the needs of applications like MVS and disparity remapping.

---

2. As described in [65], the 2D integral image is defined as $J_{xy} = \sum_{x' \leq x} \sum_{y' \leq y} I_{x'y'}$ and allows to calculate arbitrary integrals with just four memory accesses.

2.3   Algorithmic Details of SKB

When searching for sparse point correspondences in an image pair using features, three main steps have to be performed: *interest point detection, descriptor calculation* and *descriptor matching*. Next, we summarize these steps of our version of the SKB algorithm and point out the differences to the original formulation by [60]. A performance comparison with other descriptors is given at the end of this section.

2.3.1   Interest Point Detection

Similar to the original SKB implementation, we use a variant of the simple difference of boxes (DOB) filter to detect interest points in the image. However, our system builds upon the original DOB version of CenSurE [73] instead of the modified SUSurE DOB [74] that is used in the original SKB formualtion by [60]. The CenSurE detector basically performs a pixel-dense scan on all scales, whereas the SUSurE detector uses a scan line sparsification to skip pixel positions that are not likely to lead to an extremal filter response. The SUSurE detector is about $3\times$ faster than the CenSurE in software [74], but it exhibits a data dependent, irregular flow which inhibits parallel processing of different filter scales. The DOB filter is a simplified Laplacian of Gaussian (LOG) filter and its response is given by the subtraction of the pixel sums within two quadratic boxes with side length $2n + 1$ and $4n + 1$. The advantage of this simplification is that it can be efficiently calculated using an integral image [65], since the area of a box can be obtained by merely adding/subtracting the integral image values at the box corners. The image is filtered using different sizes of this filter, thereby forming a volume of filter responses which is also denoted as *scale space*. As noted in [56, 70], the largest number of interest points is found on the first few scales. Therefore, and because and in our application we do not require a large scale invariance, the scale space is limited to the first 8 scales (i.e. $n \in [1, 2, \ldots, N_{sc} = 8]$).

The DOB filter responses in the scale space are checked for extremal points in a local $3 \times 3 \times 3$ neighborhood using non-maximum suppression (NMS), and a weak response threshold t_w is applied in order to filter out non-robust interest points. Since a part of the NMS neighborhood is unknown at the scale space border, maxima are not allowed to occur there. Interest points are therefore only detected on six out of eight scales. Note, that the integrated box areas must be properly normalized [73] such that comparisons among different scales are possible. The accuracy of the interest point coordinates could be enhanced by additional interpolation of the maxima location, but since the DOB filters are evaluated pixel-dense on each scale, this is not done here. Further, no Harris corner test is performed in our implementation since this operation is very costly and often not necessary in this application [60]. The output of the interest point detection is a set of tuples (x_i, y_i, s_i) with $i \in \{0, 1, \ldots, N_{pts}\}$, where x_i and y_i are the integer image coordinates and s_i is the index of the scale of a particular interest point i.

Depending on the value of the weak response threshold t_w, different amounts of interest points are detected. It is desirable to adjust this threshold such that each frame of a video yields around the same number of points. Since the DOB filter responses are basically differences of image areas, their magnitudes correlate with the average gradient magnitude in the image. Although this relationship may be used to set the threshold to a suitable value, we found that this does not stabilize the number of points well over a video sequence since the image content itself has a large impact on how many points are detected. A feedback loop with simple rules on how to adapt the threshold has proven to be much more effective. In our version of the DOB detector, we use the following[3] rules

$$
t_w := \begin{cases}
\text{if } N_{pts} < N_{lo2} & , \ \max\left(t_w - \Delta_{w2},\ t_w^{min}\right) \\
\text{else if } N_{pts} < N_{lo1} & , \ \max\left(t_w - \Delta_{w1},\ t_w^{min}\right) \\
\text{else if } N_{pts} > N_{hi2} & , \ \min\left(t_w + \Delta_{w2},\ t_w^{max}\right) \\
\text{else if } N_{pts} > N_{hi1} & , \ \min\left(t_w + \Delta_{w1},\ t_w^{max}\right)
\end{cases} \quad , \quad (2.1)
$$

where $N_{lo2}, N_{lo1}, N_{hi1}, N_{hi2}$ are the decision boundaries, Δ_{w1}, Δ_{w2} are the step sizes, and t_w^{min}, t_w^{max} are the minimal and maximal threshold values (used for saturation). Figure 2.2a shows the behavior of the feedback loop when applied to a test video sequence containing three scene cuts. Using four decision boundaries instead of only two allows to use larger step sizes if the number of interest points is far away from the target. Note that the step size should not be chosen too large since this may cause an oscillatory behavior.

---

3. A similar functionality to control the population of detected interest points is provided by the *AdjusterAdapter* class in OpenCV [75].

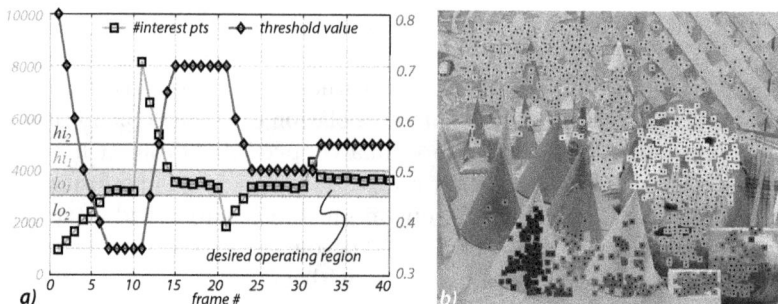

FIGURE 2.2: a) Behavior of the weak response threshold adjustment scheme in the interest point detection step (every 10 frames there is a scene change in this video sequence). The following parameter set has been used in order to stabilize the number of points between 3 k and 4 k: $N_{lo2} = 2\,\text{k}$, $N_{lo1} = 3\,\text{k}$, $N_{hi1} = 4\,\text{k}$, $N_{hi2} = 5\,\text{k}$, $\Delta_{w1} = 0.05$, $\Delta_{w2} = 0.1$ and $t_w^{min} = .1$, $t_w^{max} = 1.0$. b) example for a sparse disparity estimation using SKB 256 bit type B descriptors (estimated disparity has been color-coded). Note that there are almost no outliers even tough no post-processing step like RANSAC has been performed.

2.3.2   Descriptor Calculation

The SKB descriptor makes use of a set of sixteen 4×4 filter kernels (also called *semantic kernels*, shown in Figure 2.7b) which are evaluated at 16 positions within a normalized support region around an interest point. This leads to 256 values that are binarized using a certain thresholding scheme. Zilly et al. [60] propose three different binarization variants A, B and C where A leads to a 256 bit descriptor and B and C lead to a 512 bit descriptor. Here we use the fast variant A where the 256 values are binarized by comparing them against 0, i.e., only the sign bit is kept. Zilly et al. also define two different support regions (type A and B) out of which we use the larger region B (16×16), as our experiments show that it performs better (Figure 2.3).

The (x_i, y_i, s_i) tuples from the interest point detector are used to calculate the coordinates of the support region of that interest point. Bilinear interpolation is then used to resample the support region such that it fits into the normalized frame of 16×16 pixels (the normalization factors are given by the ratio of the outer DOB box size of the actual scale and the smallest scale i.e. $(4 \cdot s_i + 1)/5$). Note that we do not perform any rotational alignment as this is not necessary in the case of stereo matching. In order to facilitate the resampling step, we precompute an image pyramid by successive downsampling of the input image by a factor of two. Depending on the scale factor of an interest point i, the nearest pyramid level is selected as the pixel source (this concept is also known as *mipmapping* [76]). This has the advantage that aliasing artifacts are reduced in the resampled patches and that the accessed image patch is always contiguous.

2.3.3 Descriptor Matching

Generally speaking, feature matching in this context is the process of finding the optimal assignment of feature points $j \in \{1, 2, \ldots, J\}$ extracted from the left stereo image to feature points $k \in \{1, 2, \ldots, K\}$ extracted from the right stereo image. I.e., we have to solve an assignment problem of the form

$$\mathcal{S} = \arg\min_{\mathbf{S}} \sum_{j,k} (S_{jk} \cdot C_{jk}), \qquad (2.2)$$

where C_{jk} are the elements of a $J \times K$ matching cost matrix $\mathbf{C}$. $S_{jk} \in \{0, 1\}$ are the elements of a $J \times K$ assignment matrix $\mathbf{S}$ containing exactly $\min(J, K)$ nonzero elements and at most one nonzero entry per row and column. As the feature descriptors are binary vectors in this case, the matching cost between a descriptor $\mathbf{d}_j$ from the left image and a descriptor $\mathbf{d}_k$ from the right image can be efficiently calculated using the *Hamming Distance* between the two vectors, i.e.,

$$C_{jk} = \sum_{l=1}^{N_{bit}} (d_{jl} \oplus d_{kl}), \qquad (2.3)$$

where l is the bit index in the N_{bit} long vectors (256 bit in the case of SKB). However, the assignment problem itself is computationally challenging if a globally optimal solution has to be computed. E.g., the *Hungarian Method* [77] which is often employed to solve this kind of problems has a complexity of $\mathcal{O}\left(N_{desc}^3\right)$, where N_{desc} is the number of descriptors to match (assuming $N_{desc} = J = K$). Furthermore, a direct solution of the assignment problem (2.2) without any further constraints may still lead to suboptimal solutions since the geometrical relationship between the two images is completely ignored. For example the solution with the least overall cost could assign a feature point in the top right corner of the left image to a feature point in the bottom left corner of the right image, which completely violates the geometry of a rectified stereo setup. It is therefore common to constrain the search to a small window, based on the knowledge about the camera setup. In our work, we adopt a *greedy* nearest neighbour (NN) search strategy that is computationally efficient and – as will be seen in the performance evaluation in Section 2.3.4 – that provides sufficient accuracy for the problem at hand. An additional RANSAC [28] step could be performed to eliminate outliers, but this is currently not done in our HW implementation as this step is very resource consuming [64]. Also SKB combined with a small matching window provides a low outlier rate. The details of the window search are given below.

Windowed Nearest-Neighbour Search

In this work, we match points from the left image to the right, i.e., the list of descriptors from the left image is traversed, and for each point j we exhaustively search a corresponding point k in a window in the right image. This window is defined by $(\Delta_y, \Delta_{x1}, \Delta_{x2})$, and the point k has to satisfy the following set of constraints

$$\{|y_j - y_k| < \Delta_y, \quad x_k - x_j < \Delta_{x1}, \quad x_j - x_k < \Delta_{x2}\}. \qquad (2.4)$$

Δ_y is the maximum y disparity, Δ_{x1} is the maximum negative- and Δ_{x2} the maximum positive disparity. Two different Δ_x values are used here, since the positive and negative disparity ranges in a stereo setup are dependent on the alignment of the cameras and are often not equal. Δ_y accounts for disparities in y direction that occur if the image planes of the two cameras are not completely coplanar (Keystone distortion [1]). The current descriptor $\mathbf{d}_j$ from the left image is compared to all descriptors $\mathbf{d}_k$ within the matching window (2.4) using the Hamming distance. The point with descriptor $\mathbf{d}_k$ with the lowest Hamming distance is deemed the correspondence of point j if the Hamming distance lies below the matching threshold t_m (usually $\in \{10, \ldots, 64\}$ here). Once a match has been found, the points j and k are removed from the descriptor lists. This method is greedy, but it has a complexity of only $\mathcal{O}(J \cdot N_k \cdot \log(K))$, where N_k is the average number of points in the matching window (assuming

TABLE 2.1: Overview of evaluated descriptors. The size of the normalized support region (in pixels) depends on the descriptor type. Therefore, a scale calibration has to be performed for each descriptor before evaluating and comparing performance (see [78]).

	SIFT [55]	SURF [56]	BRIEF [57]	SKB-A	SKB-B
Type	Floating/Fixed Point		Binary		
Dimensions	128	64	256	256	256
Support Region	16×16	20×20	48×48	12×12	16×16
Matching Metric	ℓ^2 Distance		Hamming Distance		

that the descriptors from the right image have been sorted according to their position, and that a point lookup has complexity $\mathcal{O}\left(\log\left(K\right)\right)$.

2.3.4 Descriptor Performance

Evaluation Setup

We developed a similar framework as Mikolajczyk et al. [79] in order to compare the descriptor performance of different SKB variants against the implementations of SIFT [79], SURF [56] and BRIEF [57] (Table 2.1). The difference to the framework of Mikolajczyk et al. is that we use the Middlebury stereo test set [80], since we are interested in the performance on rectified stereo content. As explained in Section 2.3.2, the SKB descriptor comes in different flavors, out of which the computationally more efficient 256 bit types A and B are evaluated. The hardware fixed point model is included in the evaluation as well (type B). The descriptors were all evaluated on the same DOB interest points and are upright, since no orientation information is extracted in the interest point detector. The descriptor performance is assessed using *false discovery rate (FDR) vs. true-positive rate (TPR)*[4] plots (see Figure 2.3). This is a parametric plot, where the matching threshold t_m is swept from 0 to the maximum matching cost. The dataset [80] provides ground truth in the form of accurate depth maps, which were used to transform interest points from one view to the other to check for true correspondences. In this evaluation, we always used *view1* and *view5* of the scenes.

---

4. The FDR is defined as FDR = #FP/(#FP + #TP), and the TPR as TPR = #TP/(#TP + #FN), where #TP, #FP, #TN, #FN denote the numbers of true/false positives/negatives, respectively.

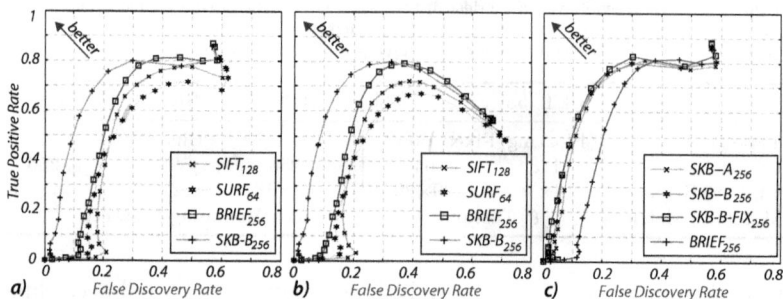

FIGURE 2.3: Matching performance simulation with the stereo test set from [80]. a) employs a globally optimal matching, b) a greedy nearest neighbour matching (left to right) and c) compares different SKB types including the fixed point model.

Results and Discussion

The evaluation results are shown in Figure 2.3. In Figure 2.3a the Hungarian method has been used to calculate a globally optimal matching, whereas in Figure 2.3b a greedy, windowed NN matching has been used. In Figure 2.3c, the two different SKB types and our fixed-point implementation are compared (the curve for BRIEF is shown as a reference, here). We can observe from Figure 2.3a and b that SKB outperforms all other descriptors, and has a very steep precision-recall curve, i.e., at low thresholds, many accurate point correspondences are found. This observation is in line with the claims of Zilly et al. [60]. When comparing the results from Figure 2.3a and Figure 2.3b we can see that a globally optimal matching performs better – especially in the region with higher thresholds. There is no 'bending down' of the precision-recall curve, as can be observed in Figure 2.3b. But we also note that in the low threshold region there is almost no difference. Since we intend to operate in this region, the greedy NN matching is a perfectly feasible choice. We can see in Figure 2.3c that among the SKB types A and B, the latter performs slightly better, and our fixed-point implementation of SKB-B shows no performance degradation.

2.4 Hardware Architecture

The complete feature extraction and matching system shown in Figure 2.1 has been developed in a series of stages. Initial work concentrated on designing an efficient *SKB core* in HW. The resulting core (described in Section 2.4.2) was implemented as a custom ASIC using a relatively mature 180 nm technology from UMC. In the later stages of the project, the development was moved to the FPGA based prototyping system TERASIC DE4-530 based on an ALTERA Stratix IV FPGA (EP4SGX530KH40C2).

The present system has been designed to be part of a much larger image processing pipeline explained in more detail in Chapter 7. The overall system adds some additional constraints on the feature extraction and matching system presented here:

1. Stereoscopic images are processed in a top-bottom format. In order to take advantage of existing video infrastructure, frames of S3D video consisting of left and right images are usually encoded as a single image in an existing video format. Two widely used formats are side-by-side, which as the name implies places the two images next to each other, and top-bottom, which places one image on top of the other.

2. A relatively high number of point correspondences (and thus many interest points) is required per frame. This is because our target application MVS [14] requires sufficiently dense depth information in the order of 5 k correspondence points to produce good results. Since the images are processed in scan-line fashion, the system must be able to cope with clusters of interest points, which demands a throughput higher than the average number of points that is detected per frame. Further, not all interest points lead to a match. So depending on the matching threshold t_m, the number of actual matches is a fraction of the amount of detected interest points.

3. One of the fundamental concerns in designing the system has been to process streaming video data and to avoid storing entire image frames within the processing core. The challenge is to balance the amount of on-chip storage with practical I/O bandwidth limitations.

In the following, we will explain individual components of the feature extraction and matching system shown in Figure 2.1, consisting of three main parts: Input buffering and image pyramid calculation, detection of interest points and calculation of the descriptors, and finally the matching unit that establishes correspondences between interest points and sends the results out via Ethernet for further processing.

2.4.1 Image Pyramid and Line Buffer

The architecture of the image pyramid calculation and the line buffer is shown in Figure 2.4a and b. The incoming 720p video is converted to gray scale and the image pyramid consisting of a 360p and a 180p resolution level[5] is calculated and stored in the off-chip RAM. The image pyramid is calculated by averaging four neighboring pixels, which results in a reduction of two in both dimensions. This averaging is implemented using two adders and a delay line per reduction step. Since the two views in the input video arrive sequentially after each other in top-bottom format, only two reduction instances are required. The calculated pixel streams are then parallelized and buffered such that long bursts can be written to the off-chip memory. In our system, the external memory is a DDR 2-6400, and a local memory controller interface provides a 256 bit wide access at 200 MHz clock rate, resulting in a peak data rate of 1.6 GB/s. The image pyramid calculation block itself works with the input pixel clock (72.5 MHz), and the off-chip bandwidth for three pyramid levels and two views amounts to 72.6 MB/s at 30 fps.

The *line buffer* keeps a sliding window of each of the pyramid levels on-chip, such that the requests from the *SKB cores* can be served with low latency and without generating additional overhead to the off-chip memory. Per view, a *DOB feed* unit transfers the lowest level of the image pyramid to the *interest point detection* stage of the *SKB cores*. The corresponding sliding window of the lowest pyramid level comprises 62 rows. Of this total, 48 rows are needed to provide the 1D integral image to the *DOB feed* in column-wise chunks (see

---

5. Since we use only eight scales in our implementation, three pyramid levels are sufficient.

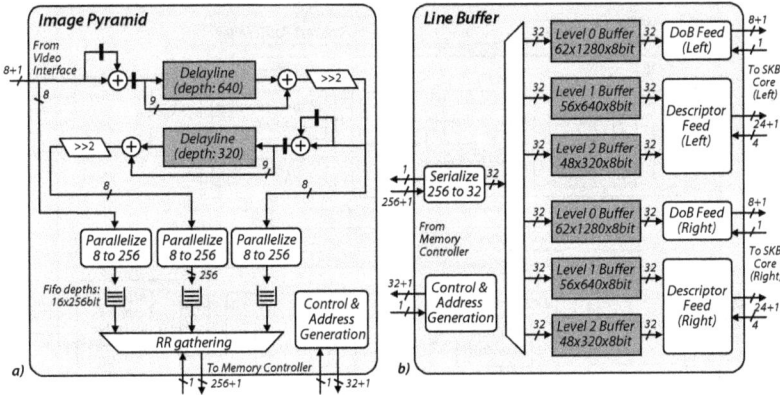

FIGURE 2.4: Details of the image pyramid a), and line buffer blocks of the feature matching system b).

Section 2.4.2). We use the remaining 14 rows for prefetching, since the cores work on overlapping image stripes of 48 rows height, and these stripes are spaced 14 pixels apart.

Two *descriptor feed* units fetch the image patches requested by the *descriptor calculation* part of the *SKB cores* from the correct image pyramid level. Out of the eight scales that are used in the interest point detection, only the scales 2 - 7 can actually have interest points due to the NMS. These scales fall onto the second and third image pyramid levels, and therefore the *descriptor feed* units only need access to these two levels. The sliding window comprises of 56 rows on the first, and 48 on the second level. This is enough to accommodate the largest descriptor that is supported (~30 pixels), plus a margin large enough such that the requested descriptor patches do not fall outside of the sliding window.

2.4.2 Detection and Extraction Core

A block diagram of the *SKB core* architecture is shown in Figure 2.5. It is composed of two main blocks which perform the interest point detection and the descriptor calculation. The *interest point detection* unit performs a dense scan over the whole image and is constantly supplied with image data by the *DOB feed* unit of the *line buffer*. The interest points detected are temporarily stored in a FIFO before

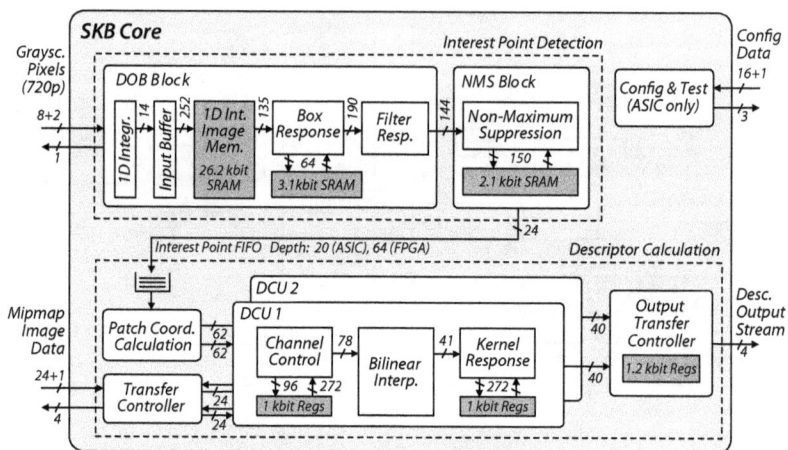

FIGURE 2.5: Block diagram of the *SKB core* with two *DCUs*: The *SKB core* is supplied with raw 8 bit gray-scale image pixels. In a first step it searches for interest points within the image. In a second step, descriptors are extracted from the surroundings of these points. For this second step, image patches around the interest points are transferred to the *SKB core*. The final descriptors are sent back to the *matching unit*.

being fetched by the *Descriptor Calculation Units* (*DCUs*). Note that the interest points are distributed sparsely over the whole image. The *DCU* has been designed to handle up to 12.5 k descriptors. Depending on the desired descriptor throughput, several instances can be operated in parallel. The FIFO serves to compensate local variations in throughput. Based on the position and scale of a certain interest point, the *DCUs* request the corresponding image patch from the *descriptor feed* units in the *line buffer*. The resulting descriptors, their position, and their scale are then sent on to the *matching unit*.

Interest Point Detection

A total of eight DOB responses have to be evaluated for each pixel (one for each scale in the scale space). Each DOB filter response can be decomposed into a linear combination of an inner and an outer box filter response. These box filter responses are usually calculated with the aid of a 2D integral image allowing to compute the sum over an arbitrary rectangular area by accessing only four values [65].

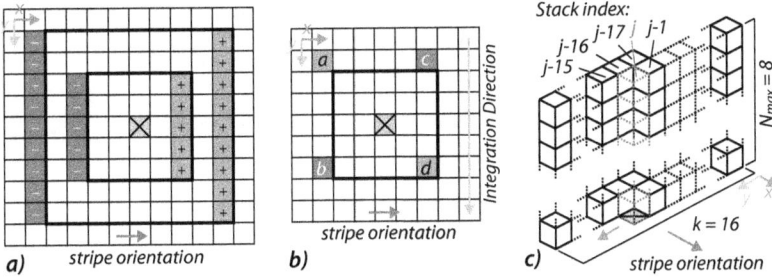

FIGURE 2.6: Interest point detection details: a) Recursive DOB filter calculation for the green pixel position. b) 1D integral image box response update. c) NMS evaluation: the most recently added stack has index j.

An integral image requires significant amounts of memory. While a conventional 720p image using 8 bit gray values requires 7.4 Mbit, the corresponding integral image requires large 28 bit entries which results in 25.8 Mbit. When the integral image is stored off-chip, this results in a large bandwidth as eight values have to be accessed per DOB filter response. For $N_{sc} = 8$ scales and an effective image size[6] of $x_{eff} \cdot y_{eff} = 1248 \cdot 688$ pixel this results in a bandwidth of $x_{eff} \cdot y_{eff} \cdot N_{sc} \cdot 8 \cdot 28$ bit/frame ≈ 1.54 Gbit/frame.

One option to reduce this bandwidth is to transfer and locally store whole blocks of the integral image in order to leverage the spatial overlap among subsequent filters [69]. In our implementation we use only a one-dimensional local integral image. Our approach builds on an observation in [74] that shows how a box filter response can be calculated recursively, provided that a dense scan is performed on the whole image. When scanning from left to right, a box filter response can be updated by adding the new pixels covered by the box on the right side, and subtracting the pixels that are no longer covered by the box on the left side (Figure 2.6a).

However, the number of additions is still linearly dependent on the filter size. In our architecture, we additionally make use of the observation that the pixel groups that have to be added or subtracted are always continuous pixel columns. It is therefore possible to use a

---

6. The effective image dimensions are given by $x_{eff} = x_{res} - 4 \cdot N_{sc}$ and $y_{eff} = y_{res} - 4 \cdot N_{sc}$, respectively.

one-dimensional integral image which enables the calculation of 1D-sums of arbitrary length along the columns in constant time (two memory accesses and one subtraction). This allows us to update a box response B_i by accessing only four corner values a, b, c, d as shown in Figure 2.6:

$$B_i = B_{i-1} - (b - a) + (d - c) = B_{i-1} + (a - c) + (d - b). \quad (2.5)$$

The terms can be reordered such that only differences between two values in the same row need to be added.

During processing, the image is scanned on all scales in parallel, as otherwise several scanning passes through the same image would be required. The *line buffer* contains sliding windows of the input images and supplies the *DOB* block in the *SKB core* with a constant stream of raw image data. The image is processed in overlapping stripes of height $h = 4 \cdot N_{sc} + N_{dob}$, where $4 \cdot N_{sc} = 32$ is the minimum neighborhood required for eight scales, and N_{dob} is the number of effectively calculated box filter responses within one column of the stripe. In order to enable non-maximum suppression in a $3 \times 3 \times 3$ neighborhood, the inner part of responses evaluated in a stripe need to be overlapped by another two pixels, i.e., subsequent stripes have a relative offset of $N_{dob} - 2$ rows. A larger value of N_{dob} reduces the overhead due to the overlap among subsequent stripes, but it also increases the size of the local integral image buffer. In our implementation we use a value of $N_{dob} = 16$. This results in a total bandwidth of $(4 \cdot N_{sc} + N_{dob}) \cdot x_{res} \cdot \left\lceil 1 + \frac{y_{eff} - N_{dob}}{N_{dob} - 2} \right\rceil \cdot 8\,\text{bit} \approx 24.1\,\text{Mbit}$ per frame, which is much lower than the original 1.54 Gbit per frame. The local integral image buffer has to store $(4 \cdot N_{sc} + N_{dob}) \cdot (4 \cdot N_{sc} + 1) = 1584$ entries.

Note, that the one-dimensional integral image can be easily constructed locally as the integration direction is orthogonal to the scanning direction. I.e., if the image is processed in stripes of a certain height h, integrating amounts adding h values, and is completely independent of the *width* of an image. This leads to a hardware architecture that only needs to store a sliding window of the original image in a *line buffer*, and the memory bandwidth can be reduced considerably compared to a naive implementation using a two-dimensional integral image.

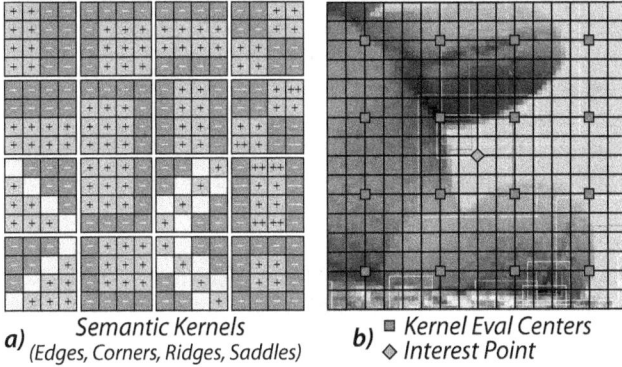

a) *Semantic Kernels*
(Edges, Corners, Ridges, Saddles)

b) ▪ *Kernel Eval Centers*
◇ *Interest Point*

FIGURE 2.7: Descriptor details: a) Descriptor support region, overlaid over an image patch. The set of semantic kernels to be evaluated on a) are depicted in b).

Descriptor Calculation

In contrast to the interest point detection part, the descriptor calculation part operates on sparse data. It has to operate fast enough to process – on average – all interest points in an image. Our evaluations have shown that several thousand interest points per frame are detected when setting the weak response threshold within a reasonable range of $t_w \in [0.1, 1.0]$. The *interest point description* block is designed to be scalable and consists of multiple parallel *DCUs* able to process up to 12.5 k descriptors per frame each. If the application requires a higher throughput, this can be easily achieved by instantiating more *DCUs* and by adjusting the bandwidth of the image memory accordingly. The current implementation uses two *DCUs* which results in an aggregated throughput of 25 k per frame.

The *interest point description* block takes interest points from the FIFO buffer and assigns each one of them to a *DCU*, which then acquires the required data from the nearest mipmap level in the *line buffer* through the *transfer controller*. Then the received data is interpolated to a normalized 16×16 image patch, followed by the evaluation of the 16 semantic kernels shown in Figure 2.7a) at 16 evaluation positions on the normalized patch (Figure 2.7b). This results in $16 \cdot 16 = 256$ responses which are then binarized via thresholding, and written to the output buffer.

The raw descriptor calculation throughput of one DCU is one descriptor in 256 cycles which translates into around 13 k descriptors per frame. However, this assumes that image data is always present for the bilinear interpolation. This is not always the case since the size of the image patches (on the mipmap levels) that are processed vary up from 16×16 to 25×25 pixels. The effective throughput is thus dependent on the speed of this interface. In our implementation, the interface can deliver 24 · 100 Mbps of pixel data, which depending on the patch size corresponds to between 15 k and 38 k image patches in the worst and best case including control overhead. This rate is sufficient to supply one DCU continuously. However, to cope with feature point clusters, it is important to have a higher throughput than what the average case suggests. This is necessary to keep the interest point FIFO size within reasonable limits. If the FIFO is too small, some of the interest points can be dropped when the density of interest points is to high. In our design, we use two $DCUs$ which together are able to process 15 k descriptors in the I/O-limited case, and up to 25 k in the computationally limited case.

2.4.3   Descriptor Matching

In this system, we are matching interest points from the left image to the right image with a the greedy, windowed nearest neighbour search explained earlier in Section 2.3.3. Since the matching process is basically an exhaustive search within the matching window, it is crucial to sort the descriptors from the right image appropriately. Otherwise, the whole memory needs to be scanned for each point from the left image. In software, this is often done by using a range-tree variant (e.g., k-d tree). Here we use a different, hardware friendlier approach similar to a direct mapped cache which has a simpler data structure management suitable for HW implementation.

Interest Point Sorting and Matching

In the worst case our right image will contain 25 k interest points, which is still very sparse and corresponds to less than 3 % of the number of pixels in a 720p image. In order to reduce both the memory and the search overhead, we have decided to use a binning

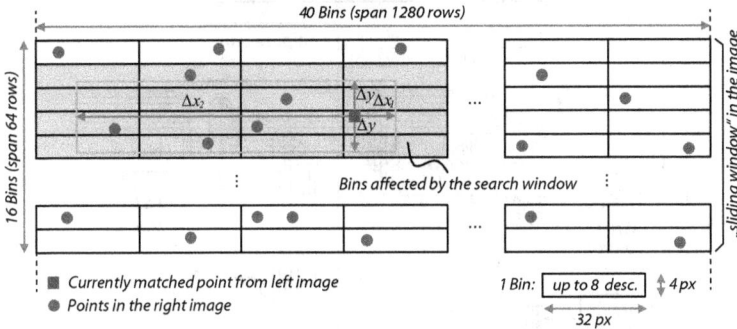

FIGURE 2.8: The image is divided into bins of 4×32 pixels. Each such bin can hold a maximum of eight interest points. Only the bins covered by the current matching search window (orange) have to be accessed.

method where we subdivide the entire image into uniformly sized bins. Each bin is allowed to store a small number of descriptors. After an exhaustive evaluation, we have decided to use a bin size of 4×32 pixels, with eight descriptor slots per bin. Assuming 25 k uniformly distributed descriptors per image, the average number of descriptors per bin is around 3.3. We use more slots per bin since sometimes clusters of interest points occur, and this may lead to a bin overflow. Our evaluations have shown that when using eight descriptor slots, the percentage of dropped descriptors is usually low and around 2 %.

In order to match a descriptor from the left image, only those bins in the right image which are covered by the matching window have to be accessed, as illustrated in Figure 2.8. When determining the size of the matching window for the nearest neighbour search, the setup of the stereo video system has to be taken into account. The exact matching window size is dependent on the geometric setup of the two cameras. Our implementation is able to handle a maximum window size of $(\Delta_y, \Delta_{x1}, \Delta_{x2}) = (15, 31, 255)$ which works fine for most setups.

Since the calculated descriptors follow a scan-line pattern from the top-left to the bottom-right of the image[7], we use a sliding-window cache arrangement. Considering the interest point detection step-size of 14 pixels and a maximum matching window height of 30 pixels, the cache should at least span $30 + 2 \cdot 14 = 58$ pixels in y dimension (14

---

7. The descriptors do not follow a strict order due to the parallel *DCU*s, and since the interest point detection works column-wise in a narrow stripe.

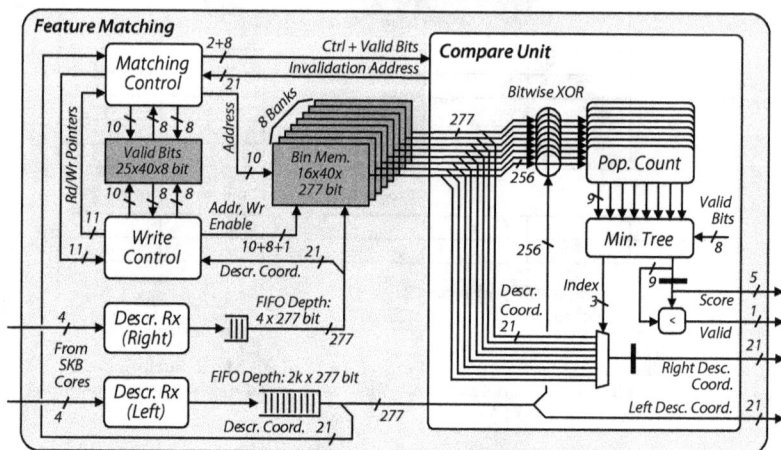

FIGURE 2.9: The *feature matching unit* contains a large descriptor FIFO for the left image, a binning memory with associated valid bits for the right image, and a *compare unit* that performs eight descriptor comparisons in parallel.

pixels to write the descriptors from the current stripe, and 14 pixels for the stripe being matched). This value is rounded up to 64 rows in our implementation.

Matching Block Details

A block diagram of the *feature matching unit* is shown in Figure 2.9. The *feature matching* unit consists of a descriptor FIFO for the left image, a binning memory for descriptors from the right image, a valid bit memory for the bins, a *compare unit* and two control units that control the binning and matching processes.

The *write control* unit contains a counter that keeps track of the sliding cache window position in the image. The coordinates of the incoming descriptors are checked against this position, and if the point lies within the sliding window, the *write control* unit checks the valid bits of the corresponding bin (there are eight valid bits per bin, one for each descriptor slot). If there is a free slot, the descriptor is written to that bin. Otherwise it is discarded. If the descriptor lies below the sliding window, the uppermost row of the sliding window is invalidated, moved to the bottom, and the row counter is incremented. This procedure is repeated until the descriptor coordinates lie within the sliding window. The *matching control* unit fetches the coordinate from the left-image descriptor currently present at the output of the FIFO, and based on the sliding window parameters, it determines which bins have to be accessed. The depth of the descriptor FIFO for the left image has been set to 2 k which is the expected number of descriptors in a 64×1280 pixel window in the image (assuming 25 k uniformly distributed descriptors). The required throughput of the *feature matching* block in terms of comparisons per second (cps) is given by the expected numbers of descriptors in the left image and in the matching window in the right image

$$25\,\text{k} \cdot 25\,\text{k} \cdot \frac{30 \cdot (31 + 255)}{720 \cdot 1280} \cdot 30\,\text{fps} = 1.74\,\text{Mcps}. \qquad (2.6)$$

Assuming a clock frequency of 100 MHz, this corresponds to a throughput of around 1.74 comparisons per cycle. But since the descriptors can be clustered, it is necessary to provide enough throughput margin such that the *feature matching* block does not start to lag behind. In our implementation, we access one whole bin in parallel and perform eight comparisons per cycle in order to stay on the safe side. These descriptors are sent to the *compare unit*, where the Hamming distances between the left image descriptor and all right image descriptors is calculated. Next, the index of the descriptor with the smallest Hamming distance is determined. After all bins have been compared, the coordinates of the descriptor pair with the smallest Hamming distance is output and the corresponding valid bits are cleared if the Hamming distance is below the matching threshold t_m.

2.5 Results

2.5.1 ASIC Implementation of the Core

As mentioned earlier, in the initial phase of the project, the *SKB core* for one view has been prototyped on an ASIC which has been fabricated in 180 nm CMOS technology (die photograph shown in Appendix A.5). Table 2.2 shows the key figures. At 100 MHz, one core is able to process a 720p video stream at 30 fps with 15 k-25 k descriptors per frame (depending on the distribution of the descriptor

TABLE 2.2: Measurement results of the SKB ASIC, and comparison with related work.

Resources	Park et al. [62]	This work
Technology	130 nm (1P6M)	180 nm (1P6M)
Core Voltage	1.2 V	1.8 V
# pads	?	82 (I:40, O:26, P:16)
Core Area	10.24 mm^2	3.08 mm^2
Logic	78.3 kGE	193 kGE
SRAM	1024 kbit	29 kbit
Total Complexity	861 kGE	254 kGE
Performance		
Descriptor	BRIEF	SKB
Functionality	ID+FE+FM[†]	ID+FE[†]
Frequency	100 MHz	100 MHz
Power (Core/Pads)	182 mW	146 mW/38 mW
Resolution	1080p	720p
Throughput	94.3 fps	30 fps
Max. Desc./Frame	512	15 k-25 k

[†] *Interest Point Detection (ID), Feature Extraction (FE), Feature Matching (FM)*

scales). As opposed to the FPGA implementation, the ASIC has a shorter interest point FIFO (20 instead of 64), does not automatically adjust the weak response threshold, and contains an additional configuration block. This block allows to adjust parameters such as the weak response threshold, and to read out statistical information on detected and dropped interest points.

For comparison, Table 2.2 also lists the specifications of the BRIEF implementation of Park et al. [62]. Note that their work has been designed for an object recognition application with different throughput requirements (1080p images, 512 descriptors per frame). Further, their implementation also contains the matching part and an on-chip descriptor buffer (4096 entries) for testing purposes.

2.5.2   FPGA Implementation of the System

The synthesis results for an ALTERA Stratix IV of our feature extraction and matching system are given in Table 2.3, together with other state-of-the-art feature extraction and matching systems. Our implementation runs with up to 142 MHz, and depending on the distribution of descriptors scales, delivers a throughput of 42 fps with 15 k-25 k descriptors per frame. Compared to the system by Wang et al. [63],

TABLE 2.3: Key figures of the FPGA implementation of our feature extraction and matching system (without memory controller, video and Ethernet interface). The pyramid calculation block runs on the pixel clock of the video interface (72.5 MHz).

Resources	Wang et al. [63]	Vourvoulakis et al. [64]	This work
Target FPGA	Virtex 5	Stratix IV	Stratix IV
LUTs	17 k	494 k	33.6 k
Registers	11.5 k	105 k	30.7 k
Memory bits	4.6 Mbit	1.89 Mbit	4.15 Mbit
DSPs	52	960	64
Performance			
Descriptor	BRIEF	SIFT	SKB
Functionality	ID+FE+FM[†]	IP+FE+FM+R[†]	ID+FE+FM[†]
Max. Frequency	159 MHz	25 MHz	142 MHz [§]
Resolution	720p	640×480	720p
Throughput	60 fps (mono)	81 fps (mono)	42 fps (stereo)
Desc./Frame	2 k	1 k	15 k-25 k

[§] Worst case corner (slow 85 °C model). [†] *Interest Point Detection (ID), Feature Extraction (FE), Feature Matching (FM), RANSAC (R)*

which implements the binary BRIEF descriptor, our system requires a similar amount of DSP slices and on-chip memory. The logic and register usage is 2× and 3× higher, respectively. This is due to the fact that we process two 720p views in parallel at 42 Hz, while Wang et al.'s system only processes one 720p view at 60 Hz. Further, our system is has about 7.5× more descriptor extraction and matching throughput.

For comparison, the recent system by Vourvoulakis et al. [64] implementing the non-binary SIFT descriptor is also listed. We can clearly observe that the binary BRIEF and SKB implementations significantly outperform SIFT in terms of resource requirements and throughput. Note however, that part of the resources of the SIFT implementation are due to the RANSAC module, which is not present in the SKB and BRIEF implementations.

2.6 Summary and Conclusions

In this chapter we looked into the design and implementation of an efficient system for the extraction of sparse point-correspondences. In particular, we take away the following points:

- We have evaluated the SKB descriptor and showed that it out-performs other state-of-the-art descriptors in terms of recognition rates in rectified stereo settings. We also demonstrate that it can be conveniently implemented in hardware, since the convolutions in the interest point detector and descriptor calculation stages can all be implemented without multipliers. In addition, due to the fact that SKB is a binary descriptor, the matching cost can be calculated in hardware-friendly manner using the Hamming distance. Considering its performance and computational simplicity, we conclude that SKB is a prime choice for efficient hardware accelerators.

- The presented system is able to extract and match SKB features from 720p stereo video in real-time, with 15 k-25 k descriptors per frame. We have used several innovations to allow on-the-fly computation, and have reduced the amount of intermediate data storage and necessary I/O bandwidth without compromising the detection quality. An important development is the dense-scan strategy for DOB interest point detection, which only operates on a local, one-dimensional integral image to compute the box filter responses. As opposed to a standard approach that employs a two-dimensional integral image, our solutions requires $3\times$ less memory in that stage. Further, the use of a local, one-dimensional integral image allows to reduce the memory bandwidth of the interest point detector by $64\times$ compared to the standard approach.

- Our system was designed for MVS and disparity remapping, where a large amount of matched interest points is needed. When comparing the system to related implementations of other binary descriptors, it is competitive in terms of resource requirements, while providing a descriptor throughput that is nearly an order of magnitude higher. Also, we show that binary descriptors such as BRIEF and SKB can be implemented much more efficiently than non-binary descriptors such as SIFT.

The presented system represents one of the main building blocks of the automatic MVS prototype presented in Chapter 7.

2.A Improved Kernels

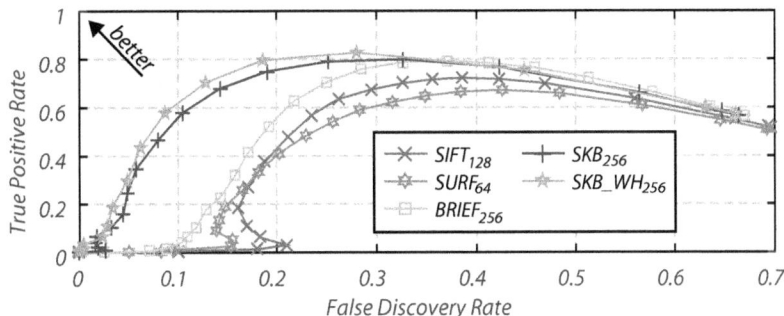

FIGURE 2.10: Performance (in terms of TPR vs. false-positive rate (FPR)) of a greedy nearest-neighbor matching with a small window, as it is implemented in hardware (test set from [80]). The 'SKB' curve is for the original 256 bit descriptor described in Chapter 2, and 'SKB_WH' represents the improved version used in Chapter 7 where the original SKB kernels have been replaced by binary Walsh-Hadamard basis functions.

During the development of the multiview system in Chapter 7, we found that SKB can be further simplified and improved by exchanging the semantic kernels shown in Figure 2.7b) by Walsh-Hadamard basis functions. This has the advantage that only the three constant factors $\{-1,0,1\}$ are needed in hardware instead of the five factors $\{-2,-1,0,1,2\}$ that were used previously. In addition, these kernels exhibit a slightly improved recognition rate as can be seen in Figure 2.10. The MVS system in Chapter 7 uses this improved variant.

Chapter 3

Spatio-Temporal Edge-Aware Filtering

3.1 Introduction

EA filters are an important tool in many image-domain applications such as HDR tone mapping [81], stylization and detail manipulation [82]. In particular, their close relation to anisotropic diffusion [83] allows to use them as efficient approximators for large, non-convex regularization problems such as optical flow, disparity estimation and colorization [84, 85]. This property becomes important considering that the size of such regularization problems typically increases rapidly with the inclusion of the temporal dimension – thereby rendering these problems often computationally impractical or even infeasible [86]. The temporal extension of EA filters is straightforward [81, 86], and computationally much more tractable than optimization-based approaches. Therefore, EA filters are well-suited for efficiently approximating many image-based regularization problems in a temporally consistent manner, which is a crucial property for video processing applications [87, 88].

However, although existing STEA methods [81, 86] are more efficient than solving complete optimization problems, they still have two major drawbacks which impede their deployment in embedded

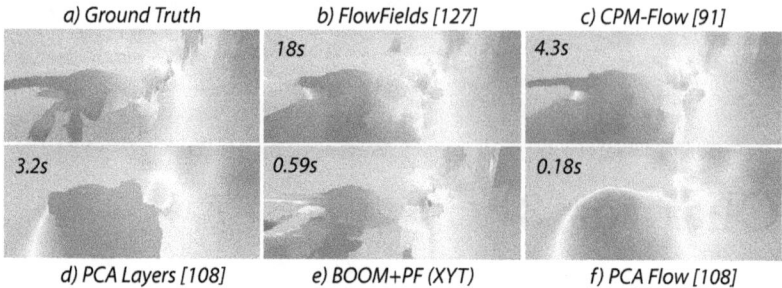

FIGURE 3.1: Optical flow calculated with fast state-of-the-art methods (b,c,d,f) and our STEA pipeline (e). Images copyright ©, BLENDER FOUNDATION, www.sintel.org.

real-time settings where only limited computational and memory resources are available. First, they either operate iteratively on complete video volumes or on sliding windows, which in both cases incurs a high processing latency and requires access to a large memory with high bandwidth. Second, the temporal filtering extension requires accurate optical flow in order to align neighboring frames within the temporal window to be filtered, which is difficult to obtain efficiently. In this chapter, we address these issues by proposing an efficient STEA filtering pipeline based on the recently proposed permeability filter (PF) [81, 89]. In particular, we make the following contributions:

- We show that under certain conditions, the temporal PF can be formulated as an IIR filter in time, which reduces computational complexity and bandwidth requirements compared to windowed approaches, since the filter can be incrementally evaluated over time. Also, this formulation exhibits constant memory complexity irrespective of the actual video length.

- In order to get fast and accurate flow estimates, we design a new binary feature descriptor termed BOOM which outperforms other state-of-the-art descriptors [44, 55–57, 59, 60, 90] in terms of receiver operating characteristic (ROC) performance. We integrate it into the recently proposed coarse-to-fine patch match (CPM) method [91] in order to efficiently compute a quasi-dense nearest neighbour field (NNF) [92] which is used as an optical flow initialization. Dense flow is then obtained via interpolation with the PF.

- Combining the above two components, we design an efficient STEA filtering pipeline and provide an efficient implementation that is significantly faster than previous methods [81, 86]. Our single-core CPU implementation processes an MPI Sintel frame [93] in only 0.59 s (resulting in a throughput of 1.3 MPixel/s) and therefore paves the way for embedded implementations in resource-constrained real-time settings.

- We present results for several image-based applications such as optical flow, disparity and saliency estimation. In particular, we demonstrate that our implementation significantly outperforms other state-of-the-art optical flow methods in terms of both speed and quality (see Figure 3.1).

After an introduction of related work in Section 3.2 we explain our STEA pipeline in Section 3.3 and the CPM+BOOM method in Section 3.4. The results and comparisons are presented in Section 3.5, and Section 3.6 finally concludes this chapter.

3.2 Related Work

3.2.1 Edge-Aware Filtering

EA filters are important basic building blocks in many image and video processing methods. Milanfar et al. [94] give an extensive overview of many filtering approaches. Since the establishment of the first EA methods such as the bilateral filter [95], a variety of EA filters have been proposed [81, 82, 96–101] for several image-based applications such as stylization, HDR tone mapping, detail editing and noise reduction. Notable EA filters are the following:

- The so-called weighted least squares (WLS) filter [96] is often used as reference method due to the high quality of its results and its ability to suppress halo artifacts by penalizing the distance between the original and filtered image. However, the drawback of this method is its high computational cost, since it requires the solution of a large linear system.
- Edge-avoiding wavelets [97] are simpler to compute, but suffer from aliasing problems and irregular edges [101].

- The local Laplacian filter is capable of producing high-quality results, but it also has the downside of being computationally demanding [101]. The authors of [100] provide a newer, more efficient implementation of the Laplacian filter, but it is not clear how to extend the method to the temporal domain.
- Other recent EA filtering techniques such as the guided filter (GF) [98], the PF [81,89], the domain transform (DT) [82] and its extension for high-order recursive filtering [99] are all very efficient techniques which offer a very good quality-performance tradeoff. However, as pointed out by Aubry et al. [100], the GF and DT still suffer from halo-artifacts. The PF does not suffer from this issue, since it has specifically been designed to mimic similar behaviour as the high-quality WLS filter – but with significantly lower computational complexity.

3.2.2 Optimization Problems

A particularly interesting aspect of EA filters is their close relation to *anisotropic diffusion* [83, 102] which recently spurred a trend to use image filtering techniques to approximate optimization problems [84–86]. As explained in detail in [86], EA filters can be leveraged for a class of regularization problems which minimize energy functionals of the form $E(\mathbf{f}) = E_{data}(\mathbf{f}) + \lambda E_{smooth}(\mathbf{f})$, where E_{data} is the application specific error term, E_{smooth} enforces smoothness among neighbouring pixels and $\mathbf{f}$ is an unknown solution. Rather than imposing a regularization term E_{smooth} and solving for $\mathbf{f}$, smoothness can be created by filtering application-specific initial conditions (which minimize E_{data} locally) with an efficient EA operation. By implicitly calculating optical flow, Lang et al. [86] also extend this concept to the temporal dimension and propose an iterative STEA filtering method which is able to efficiently calculate *temporally stable* results for a variety of applications ranging from optical flow estimation itself to disparity and visual saliency estimation, sparse data upsampling and scribble propagation.

3.2.3 Temporal Consistency

Temporal consistency is a significant problem in video processing since the frame-by-frame application of image-based methods can often produce visually disturbing temporal artifacts [81, 86–88]. As mentioned in [88], STEA filters [81, 86] are good at removing high-frequency temporal artifacts such as noise and flickering, but for low-frequency instabilities, more elaborate methods such as [87] and [88] have to be used. These algorithms can deliver better quality for certain applications - but at the cost of significantly increased complexity. Both [87] and [88] need to solve large optimization problems and typically require precomputed, accurate optical flow. Despite the limitation of STEA filtering methods mentioned, they provide efficient means for introducing temporal consistency and can be used to bootstrap accurate optical flow using sparse correspondences and interpolation. Our work is conceptually similar to [86], but focuses on addressing the following two issues which are critical for embedded real-time settings, where only limited computational and memory resources are available.

First, we formulate a temporal filtering extension which does not require access to a large memory with high bandwidth. In particular, instead of using the DT filtering kernel [82], we chose to use the PF kernel instead, that has been originally introduced in [89] to filter disparity data, and which has been successfully extended to the temporal domain to filter HDR data in [81]. Under the simplifying assumption to only filter over all past video frames, the kernel it can be reformulated as a non-linear IIR filter in time. This obviates the need to iterate over complete video volumes and results in constant computational complexity and memory requirements. The PF has similar spreading characteristics as the DT, which is important for propagation of sparse data (e.g., sparse data upsampling). In addition, the PF can be implemented with linear complexity $\mathcal{O}(N)$, where N is the amount of processed pixels, and the filter has good halo-reduction capabilities due to its close relation with the high-quality WLS filter.

Second, we propose a more accurate and efficient way to obtain optical flow estimates. Lang et al. [86] compute sparse point correspondences using standard SIFT features which are then fed into their filtering framework to iteratively compute dense optical flow for the temporal filter. These standard features suffer from the fact that they are usually not regularly spaced, but rather occur in clusters in textured areas. In order to obtain sufficient coverage of the whole image, the interest point detectors therefore have to be tuned to yield many more points, which makes the matching process slow and inefficient. In this work, we also use the filtering framework to interpolate a sparse flow-initialization, but we propose to use a more accurate and efficient matching method, which returns a regular, quasi-dense correspondence field.

3.2.4 Optical Flow Estimation

Optical flow estimation is an extensive field and therefore we only mention relevant methods and refer to [103, 104] for more details on standard algorithms.

Several recently proposed optical flow methods rely on sparse feature matching methods in order to gain efficiency. Sparse feature matches obtained between two temporally adjacent frames are typically used as initializations for variational refinements [91, 105, 106] or interpolation methods [86, 107, 108]. In particular, a recently proposed multi-level method called CPM [91] has been shown to provide a good speed/performance tradeoff (currently it is among the top ranking methods on the MPI Sintel dataset [93]). CPM calculates a quasi-dense NNF via a randomized nearest neighbor search. Compared to standard feature matching approaches which first employ an interest point detector, such NNFs achieve much more regularly distributed point correspondences. However, standard NNFs such as computed by PatchMatch [92] originally suffered from many outliers. This issue has been addressed by CPM by performing the randomized search on an image pyramid and making use of SIFT descriptors [55] instead of a block matcher. In order to gain efficiency, the NNF is calculated on a subsampled grid – and hence it is quasi-dense. In order to obtain a dense flow estimate, the NNF is interpolated using the geodesic interpolation and variational refinement method from EpicFlow [106]. CPM-Flow takes around 4.3 s on a single core machine, where around 1.3 s are spent calculating SIFT descriptors and performing CPM, and the remaining 3 s are spent in the variational refinement step of EpicFlow. Note that CPM-Flow has similarities with SIFT-Flow [109] and DAISY [90], which are both methods that use a dense scan descriptor formulation in order to efficiently compute a descriptor for each pixel. In fact, CPM-Flow uses the same dense formulation as SIFT-Flow to efficiently calculate the descriptor field.

In this work, we leverage the advantages of CPM, but without the costly SIFT descriptors and the variational refinement step. To this end, we propose to use CPM with an efficient, binary descriptor which we specifically developed for this task. The variational refinement step is then approximated using our much faster EA filtering pipeline.

3.2.5 Features

Current state-of-the-art descriptors can be categorized into two classes, namely *floating-point* methods such as SIFT [55], SURF [56] or DAISY [90] which produce numeric feature vectors and *binary* methods which produce bit vectors. The latter class of descriptors has received increased attention over the last couple of years since they often can be calculated and matched more efficiently than their floating-point counterparts. Notable binary standard methods are BRIEF [57], BRISK [58], FREAK [59] and SKB (Chapter 2). Zhou et. al. [110] propose a binary version of SIFT called B-SIFT that cannot be extracted as efficiently as the other binary descriptors mentioned, since it computes the binary representation from standard SIFT descriptors in an additional post-processing step.

An interesting observation is that gradient-based descriptors like histogram of oriented gradients (HOG) [111] (and variants thereof) often exhibit superior ROC performance compared to intensity-based descriptors [112–114]. Moreover, HOG-based features have also proven to be useful for calculating optical flow with large displacements [115].

Machine learning has been successfully applied to optimize feature arrangements and reduce the dimensionality of the vectors [78, 112–114, 116]. In particular, methods employing *AdaBoost*[1] have recently led to a variety of top-performing descriptors such as BinBoost [112], Bamboo [113] and LDDB [114]. However, learned descriptors typically forfeit some of the efficiency gain when extracting the descriptor, since they often have irregular computation patterns and employ costly dimensionality reducing projections.

For the NNF calculation using CPM, a good tradeoff among extraction speed, matching speed and ROC performance is important. On one hand we have state-of-the-art methods which provide fast feature extraction, but exhibit a larger memory footprint and fair ROC performance. On the other hand, there exist descriptors with much smaller memory footprint and/or better ROC performance – but these are typically more costly to compute. We strive to seek a balance between these attributes and present a simple binary descriptor termed BOOM, that can be efficiently calculated and matched, and exhibits similar performance as SIFT and DAISY (see comparison in Section 3.5). Our descriptor is inspired by HOG-like methods, but takes them a step further by providing an effective way to obtain a compact 256 bit representation. It should be noted that our descriptor has similarities with binarized histogram of oriented gradients (BHOG) variants [117–119] from the domains of human detection and sketch-based image retrieval. However, as described in more detail in Section 3.4.1, our descriptor employs a different cell layout and binarization strategy, and is based on orientation maps introduced by Tola et al. [90] instead of standard HOGs. These design decisions lead to significantly better ROC performance as demonstrated in Section 3.5.

3.3 STEA Filtering Pipeline

An overview of our STEA filtering pipeline is shown in Figure 3.2. The inputs are a *guiding video sequence* $\mathbf{I}_t$ and *additional attribute channels* to be filtered $\mathbf{A}_t$, where t is the frame index. The guiding frames $\mathbf{I}_t$ are used to derive the edge-aware filter coefficients[2] $\mathbf{H}^{XY}$ and to estimate sparse optical flow $\mathbf{F}_t$. The filter coefficients $\mathbf{H}^{XY}$ are then used to perform spatial filtering operations in order to turn the sparse optical flow into a dense flow field $\mathbf{F}_t^{XY}$ and to filter the additional data channels in spatial directions to produce $\mathbf{A}_t^{XY}$. The

---

1. AdaBoost is a learning technique used to select and combine weak classifiers into a more powerful ensemble classifier. One of the well known classifiers trained using this technique is the face detector by Viola and Jones [65].
2. It should however be noted that the efficient scan line formulation summarized in Section 3.3.3 does not explicitly calculate $\mathbf{H}^{XY}$.

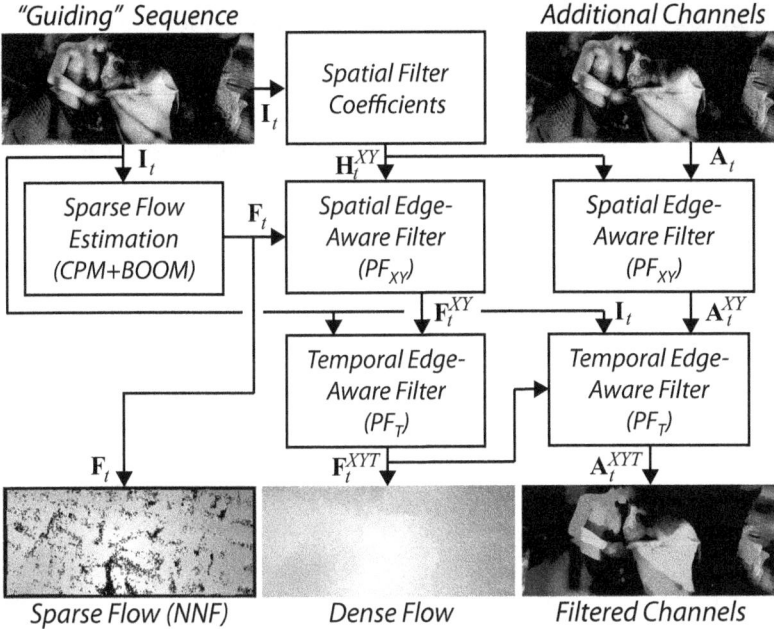

FIGURE 3.2: Overview of the STEA Filtering Pipeline. Refer to the text for more details. Images copyright ©, BLENDER FOUNDATION, www.sintel.org.

optical flow estimate $\mathbf{F}_t^{XY}$ is then used to enable temporal filtering along motion paths. First, it is used to temporally filter itself to produce a stabilized version $\mathbf{F}_t^{XYT}$, and this temporally stabilized flow is then used to filter the additional channels in time. In this example we have $\mathbf{A}_t = \mathbf{I}_t$ in order to illustrate the impact of EA filtering, but in practice these usually contain other feature maps such as disparity or saliency values. The pipeline outputs sparse and dense flow estimates $\mathbf{F}_t$, $\mathbf{F}_t^{XYT}$, as well as the filtered channels $\mathbf{A}_t^{XYT}$.

Similar to other state-of-the-art flow estimation algorithms [91, 105–107], flow is estimated using sparse feature extraction, followed by an efficient sparse-to-dense conversion. The difference of our method w.r.t. related work is, that our conversion consists of a fast edge-aware interpolation without any variational refinement. We will see in Section 3.5, that this approach offers a competitive speed-quality trade-off. The employed feature matching method (CPM+BOOM) is an improved NNF method based on sparse features and will be explained in Section 3.4.

For the EA filtering operation we selected the recently proposed PF [81, 89] due to its high efficiency and good quality. It can be formulated as an iterative application of ideally parallelizable 1D filtering passes with constant complexity per processed pixel. This property is shared by the DT filter [82], but the PF has the additional benefit of being closely related to the WLS filter, which is known to produce high-quality anisotropic filtering results with good halo artifact reduction [82, 96, 97]. The temporal filter employs the same PF, but as opposed to its original extension to the temporal domain [81], we formulate it as an incremental IIR filter which does not iterate over large temporal volumes. This enables efficient, low-latency implementations with constant memory complexity.

In the following, we will describe the spatial and temporal filtering steps, and the employed NNF method for flow computation is explained in the subsequent section.

3.3.1   Spatial Filtering of Dense Data

The filter used in this work belongs to a class of filters defined by iterative application of the recursion

$$J_p^{(k+1)} = \sum_{q \in \mathcal{D}_{in}} H_{pq} J_q^{(k)} + \lambda^{XY} H_{pp} \left(A_p - J_p^{(k)} \right), \qquad (3.1)$$

where A_p denotes the input data to be filtered at position p at frame t, $J_p^{(k)}$ is the diffusion result at position p after k iterations. The set $\mathcal{D}_{in}$ contains all pixel positions of a frame, and H_{pq} are elements of the row stochastic [94] matrix $\mathbf{H}$, that defines the filter. The iteration is initialized with $\mathbf{J}^{(0)} = \mathbf{A}$. Frame indices are omitted in this subsection since all operations are applied to spatial dimensions exclusively. The first term of (3.1) is the actual shift-variant convolution, and the second term is a fidelity term where $\lambda^{XY} \in [0, 1]$ can be used to bias the iteration towards the input data $\mathbf{A}$. Aydın et al. [81] explain that the choice of $\lambda^{XY} = 1$ significantly reduces halo artifacts.

The PF is a specific instance of (3.1) with two separate filter matrices $\mathbf{H}^X$ and $\mathbf{H}^Y$ for filtering operations in horizontal and vertical

Guiding Image **I**   Permeability Maps $\tilde{\pi}^X$ and $\tilde{\pi}^Y$

Channels **A** $\mathbf{A}^{XY}$ after 1 and 5 iterations

FIGURE 3.3: Edge-aware spatial filtering using permeability maps $\tilde{\pi}^X$ and $\tilde{\pi}^Y$. Note that **A** is set to **I** for illustrative purposes in this example. Images copyright ©, BLENDER FOUNDATION, www.sintel.org.

direction, respectively. These operations are applied in alternating fashion, and the concatenation of one X and one Y pass constitutes one spatial filter iteration. The two matrices $\mathbf{H}^X$ and $\mathbf{H}^Y$ are defined via *permeability weights*[3] π_{pq} between two pixels p and q which control the local diffusion strength. Below we summarize how ro obtain the coefficients $\mathbf{H}^X$ for the horizontal filter. The derivation for $\mathbf{H}^Y$ is analogous. The permeability between two neighboring pixels $p = (x, y)$ and $p' = (x + 1, y)$ is defined as

---

3. The following list summarizes a couple of image-based quantities used:

   Permeabilities: Permeabilites are used to control the local diffusion strength of the PF. They are derived from the input image according to (3.4), and capture how *permeable* the image is between two particular pixels. The permeabilities of a frame form a permeability map, and there exists a map for the horizontal and for the vertical direction (denoted as π^X and π^Y).

   AEE: The average endpoint error (AEE) is given by

   $$AEE\,(\mathbf{F}, \mathbf{T}) = 1/\,|\mathcal{D}_i| \cdot \sum_{k \in \mathcal{D}_{in}} \|\mathbf{f}_k - \mathbf{t}_k\|, \qquad (3.2)$$

   and measures the average error distance between a two-dimensional optical flow field $\mathbf{F}$ and the ground truth $\mathbf{T}$.

   Photo Constancy: The permeability map used to guide the PF in temporal dimension is denoted as photo constancy, and can be calculated with (3.13).

   Gradient Measure: The gradient measure is calculated according to (3.14), and captures rapid optical flow changes based on the flow gradients. It is used in addition to the photo constancy in order to reduce filter and warping artifacts in image regions with complex motion.

$$\pi_{\tilde{\pi}}^{X} = \left(1 + \left|\frac{\|I_p - I_{p'}\|}{\sqrt{3} \cdot \sigma^{XY}}\right|^{\alpha^{XY}}\right)^{-1}, \tag{3.3}$$

which is a variant of the Lorentzian edge-stopping function, applied to the color distance between p and p' *of the guiding image* **I**. This function evaluates close to zero if the color distance between these two pixels is high, and to one if the difference is low. The parameter σ^{XY} controls the transition point and α^{XY} the falloff rate. Typical values are $\sigma^{XY} \approx 0.02$ and $\alpha^{XY} = 2$. Permeabilities between arbitrary pixels are then defined as

$$\pi_{pq}^{X} = \begin{cases} 1 & \text{if } p = q, \\ \Pi_{n=p_x}^{q_x-1} \tilde{\pi}_{(n,p_y)}^{X} & \text{if } p_x < q_x, p_y = q_y \\ \Pi_{n=q_x}^{p_x-1} \tilde{\pi}_{(n,p_y)}^{X} & \text{if } p_x > q_x, p_y = q_y \\ 0 & \text{else.} \end{cases} \tag{3.4}$$

The final filter coefficients h_{pq} are then obtained by normalizing the pairwise permeabilities as

$$H_{pq} = \pi_{pq}^{X} \left(\sum_{n=1}^{w} \pi_{(n,p_y),q}^{X}\right)^{-1}, \tag{3.5}$$

where w is the image width. This filtering process is illustrated in Figure 3.3, where the guiding image is also subject to filtering. We observe that even with a low number of XY passes in the order of five iterations, the PF achieves strong edge-aware diffusion. Note, that the permeabilities in (3.4) are defined such that the filtering operations reduce to 1D operations over image rows or columns. As will be shown later, this filter can be implemented with efficient scanline operations.

3.3.2   Spatial Filtering of Sparse Data

Although most EA filters are not strictly interpolating filters, they can also be used to efficiently spread sparse data, i.e., to perform an edge-aware, sparse-to-dense conversion. As shown in [86], this can be conveniently achieved by introducing a normalization map **G**, that contains nonzero values at sparse sample positions, and is zero otherwise. The map **G** is subject to the same filtering operation which is applied to the corresponding sparse data channels. After the desired

Sparse Flow **F**   After 5 Iterations

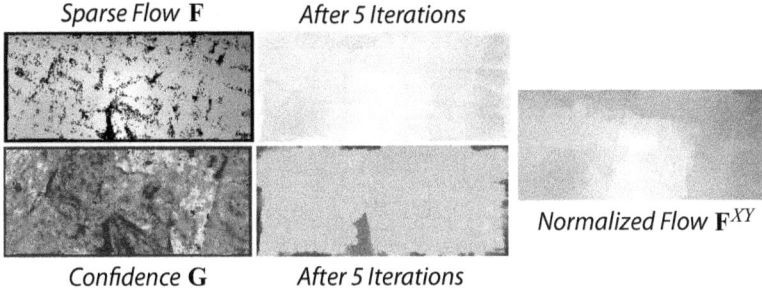

Normalized Flow $\mathbf{F}^{XY}$

Confidence **G**   After 5 Iterations

FIGURE 3.4: Edge-aware spatial filtering of sparse data (in this case 2D flow vectors). Images copyright ©, BLENDER FOUNDATION, www.sintel.org.

amount of filtering iterations k, the map is used to normalize the filtered data **F** element-wise as

$$\mathbf{F}^{XY} = \mathbf{F}^{(k)}./\mathbf{G}^{(k)}, \tag{3.6}$$

where the operator $./$ denotes an element-wise division, and **F** denotes the sparse data channels to be filtered. Figure 3.4 illustrates the sparse-to-dense conversion of optical flow. Note, that the normalization map **G** can additionally be used to incorporate *data confidence* by assigning values between 0.0 and 1.0 at the sparse sampling positions in order to give more weight to those samples which are considered to be more accurate than others. For sparse flow-vectors, we use the matching confidence, normalized to the range $[0.0, 1.0]$.

3.3.3 Efficient Formulation of the Spatial Filter

As shown by [89], the multiplicative concatenation of the permeabilities allows to formulate the filtering operation as an efficient two-pass scan line operation with constant computational complexity per pixel. The formulas are only given for the k-th horizontal iteration since their counterparts for the vertical iteration follow analogously.

As illustrated in Figure 3.5, the intermediate results l_p and the corresponding normalization values $\hat{l}_p$ are computed in a first left-right scan line pass using the recursions

$$
\begin{aligned}
l_p &= \tilde{\pi}^X_{(p_x-1,p_y)} \left(l_{(p_x-1,p_y)} + J^{(k)}_{(p_x-1,p_y)} \right), \\
\hat{l}_p &= \tilde{\pi}^X_{(p_x-1,p_y)} \left(\hat{l}_{(p_x-1,p_y)} + 1.0 \right).
\end{aligned}
\tag{3.7}
$$

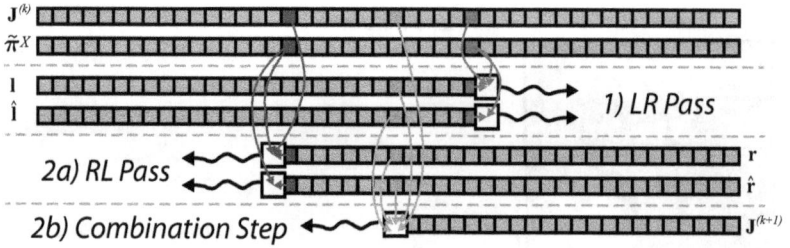

FIGURE 3.5: The PF can be efficiently evaluated with only two scan line passes due to the multiplicative concatenation of the permeabilities. As explained in the text, left- and right-sided intermediate results can be formed, which allow to compute the final result without explicitly evaluating the full convolutions.

In a second right-left pass (2a in Figure 3.5), the right-sided quantities r_p and $\hat{r}_p$ are then computed as

$$r_p = \tilde{\pi}_p^X \left(r_{(p_x+1,p_y)} + J_{(p_x+1,p_y)}^{(k)} \right),$$
$$\hat{r}_p = \tilde{\pi}_p^X \left(\hat{r}_{(p_x+1,p_y)} + 1.0 \right). \tag{3.8}$$

The result is finally calculated by combining and normalizing the intermediate results (step 2b in Figure 3.5), and adding the bias term $\lambda^{XY} \cdot \left(A_p - J_p^{(k)} \right)$ as

$$J_p^{(k+1)} = \frac{l_p + \left(1 - \lambda^{XY} \right) \cdot J_p^{(k)} + \lambda^{XY} \cdot A_p + r_p}{\hat{l}_p + 1.0 + \hat{r}_p}. \tag{3.9}$$

Note, that this third step can be efficiently carried out on-the-fly during the right-left pass, since all intermediate results are available at position p at this point. Therefore, the whole procedure results in exactly two scan line passes. Individual scan lines of one X or Y iteration are independent and can be conveniently parallelized. The initial values $l_{(1,p_y)}$, $\hat{l}_{(1,p_y)}$, $r_{(w,p_y)}$, $\hat{r}_{(w,p_y)}$ are all set to zero (w is the image width).

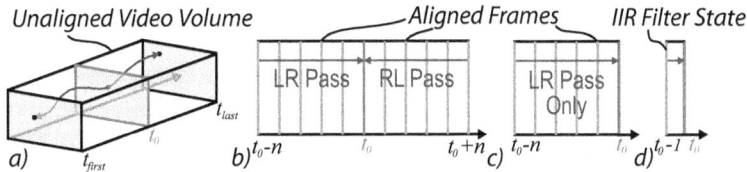

FIGURE 3.6: Different temporal extensions of the filter: a) iterative filtering along motion paths within the complete video cube, b) single filtering step within aligned symmetric (noncausal) window, c) single filtering step within left-sided (causal) window, d) IIR formulation of c).

3.3.4 Temporal Filtering

Existing Approaches

In related STEA filtering methods [81, 86], the temporal extension of the EA filter has been achieved in two different ways, and both solutions can be problematic in terms of computational- and memory complexity when targeting efficient STEA embodiments capable of real-time operation. Lang et al. [86] iteratively apply XY and T iterations to the complete video volume by following the motion paths as illustrated in Figure 3.6a. Similarly to our work, they use their DT-based filter to bootstrap optical flow. Their method works on a large data structure of linked lists representing the motion paths, and requires a large amount of high-bandwidth memory in order to store the complete video volume. Also, the approach is limited to batch-wise processing, which incurs a high processing latency and therefore makes this approach unsuitable for real-time, stream processing settings. Aydın et al. [81] formulate their filter on a reduced temporal sliding window comprising in the order of ±10 spatially aligned frames, which reduces the latency to about ten frames. Also, they reduce the computational complexity by applying only one T iteration after the spatial XY iterations. This turned out to be sufficient to obtain temporally smooth results. However, the need to align all frames within the temporal window still incurs a non-negligible computational overhead. We therefore refrain from these approaches and seek to formulate the filter in an incremental fashion, which can be evaluated very efficiently and with constant memory complexity. As shown in more detail below, this is possible by leveraging the mathematical properties of the permeability filter.

Formulation of the Recursive Permeability Filter

First, we make the following two assumptions:

- We assume that only one T iteration is applied after the XY iterations. This assumption is valid in practice since a single T iteration improves temporal consistency [81]. In addition, using just one T iteration eliminates the 'chicken-and-egg' problem described in [86], which arises when the dense optical flow estimate used in the frame alignment is bootstrapped using the filter itself.
- We assume that all data to be filtered, $\mathbf{J}_t^{XY}$, has been aligned to the centering frame t_0 within a certain temporal neighbourhood $\mathcal{T} = [t_0 - n, ..., t_0, ...t_0 + n]$, $n \in \mathbb{N}$. For the moment we neglect the fact that this assumption implicitly requires the availability of optical flow for these frames in order to perform the alignment, as this does not pose a problem anymore in the incremental formulation.

With these assumptions, we can calculate one T filtering iteration with the same recursion equations defined in (3.7) and (3.8), but using temporal permeabilities $\tilde{\pi}_t^T$ which will be defined in Section 3.3.5. In other words, we can compute the intermediate results $\mathbf{l}_t$, $\hat{\mathbf{l}}_t$, $\mathbf{r}_t$, $\hat{\mathbf{r}}_t$, where a left-right pass corresponds to a forward pass in time and vice versa for the right-left pass (the bold symbols now represent full-frame matrices, indexed by the frame number t). The result of one T iteration for the frame at t_0 is then given by element-wise division

$$\mathbf{J}_{t_0}^{XYT} = \frac{\left(\mathbf{l}_{t_0} + \left(1 - \lambda^T\right) \cdot \mathbf{J}_{t_0}^{XY} + \lambda^T \cdot \mathbf{A}_{t0} + \mathbf{r}_{t_0}\right)}{\left(\hat{\mathbf{l}}_{t_0} + 1.0 + \hat{\mathbf{r}}_{t_0}\right)}. \tag{3.10}$$

We now simplify the problem by replacing the symmetric temporal neighborhood with a causal, one-sided time window $\mathcal{T} = [t_0 - n, ..., t_0]$, $n \in \mathbb{N}$. This is a valid simplification, especially for real-time settings where low-latency is required and no information about future frames is available. The recursion equation then reduces to

$$\mathbf{J}_{t_0}^{XYT} = \frac{\left(\mathbf{l}_{t_0} + \left(1 - \lambda^T\right) \cdot \mathbf{J}_{t_0}^{XY} + \lambda^T \cdot \mathbf{A}_{t0}\right)}{\left(\hat{\mathbf{l}}_{t_0} + 1.0\right)}, \tag{3.11}$$

since r_{t_0} and $\hat{r}_{t_0}$ are zero in this case. This means that one temporal iteration can basically be calculated using just one left-right pass. If we now let $n \to \infty$, and by considering the fact that the left-right pass is defined as a recursion, we see that it is possible to obtain a non-linear IIR filter in time which only requires one recursion for each time step. The only missing part to consider is alignment. Recall that we assumed that all frames within $\mathcal{T}$ are aligned to the frame t_0 in the first place. If we drop this assumption and want to reuse the IIR filter state of the previous time-step $t_0 - 1$ to update the recursion equation, we have to re-align it to the current frame t_0, which can be conveniently done using forward-warping based on the flow estimate of $\mathbf{F}_{t_0-1}^{XYT}$ which is already computed and available:

$$
\begin{aligned}
\mathbf{l}_{t_0} &= \tilde{\boldsymbol{\pi}}_{t_0}^T \, \mathrm{warp}_{\mathbf{F}_{t_0-1}^{XYT}} \left(\mathbf{l}_{t_0-1} + \mathbf{J}_{t_0-1}^{XY} \right), \\
\hat{\mathbf{l}}_{t_0} &= \tilde{\boldsymbol{\pi}}_{t_0}^T \, \mathrm{warp}_{\mathbf{F}_{t_0-1}^{XYT}} \left(\hat{\mathbf{l}}_{t_0-1} + 1.0 \right).
\end{aligned}
\tag{3.12}
$$

Note, that this recursion step in time can be efficiently implemented with constant memory and low latency. For the implementation of the forward mapping operator $\mathrm{warp}_{\mathbf{F}_{t_0-1}^{XYT}}(.)$, we use the EWA splatting framework [32, 51, 120] which is an efficient high-quality resampling technique for non-linear image transformations.

3.3.5 Temporal Permeabilities and Flow Constancy

As proposed in [81], we use a combination of color constancy and flow-gradient magnitude in order to calculate permeabilities in temporal direction. The photo constancy is a straightforward extension of the spatial permeabilities

$$
\tilde{\boldsymbol{\pi}}_t^{photo} = \left(1 + \left| \frac{\left\| \mathbf{I}_t - \mathrm{warp}_{\mathbf{F}_{t-1}^{XYT}} (\mathbf{I}_{t-1}) \right\|}{\sqrt{3} \cdot \sigma^{photo}} \right|^{\alpha^{photo}} \right)^{-1},
\tag{3.13}
$$

and allows filtering along motion paths with similar color values. The gradient-magnitude measure is calculated similarly as

$$\tilde{\pi}_t^{grad} = \left(1 + \left| \frac{\left\| \mathbf{F}_t^{XY} - \text{warp}_{\mathbf{F}_{t-1}^{XYT}} \left(\mathbf{F}_{t-1}^{XYT} \right) \right\|}{\sqrt{2} \cdot \sigma^{grad}} \right|^{\alpha^{grad}} \right)^{-1}, \quad (3.14)$$

and prevents temporal filtering with complex motion where flow and warping artifacts are likely to occur. In other words, this measure introduces a bias towards the current frame in regions where temporal filtering may introduce errors. Note that division and exponentiations are all element-wise in the two equations above.

The two measures are then multiplied element-wise to get the final temporal permeabilities $\tilde{\pi}_t^T = \tilde{\pi}_t^{photo} \cdot \tilde{\pi}_t^{grad}$. Figure 3.7 shows two examples, where the first shot exhibits slow and consistent camera zoom, and the second shot contains quickly moving parts (left arm of the main character). We can see that the gradient measure effectively detects regions where warping artifacts are likely to occur.

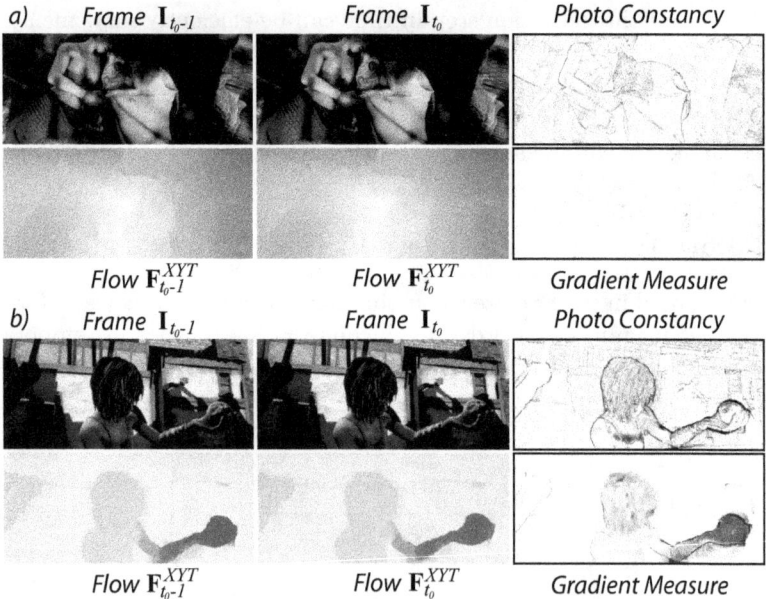

FIGURE 3.7: Photo constancy and gradient measures for slow and steady zoom (a), and fast-changing motion (b). Images copyright ©, BLENDER FOUNDATION, www.sintel.org.

3.4 CPM with Binary Descriptors

Our STEA pipeline relies on sparse flow vectors to obtain dense optical flow via EA filtering. These flow vectors are extracted using an improved version of the recently proposed CPM [91] method, which we will explain in the following.

CPM is a NNF method, developed to provide accurate optical flow vectors on a coarse, but regular grid. As opposed to the original PatchMatch algorithm [92], the randomized search is formulated on subsampled grids over an image pyramid, where matching information is propagated from coarser to finer grids in a top-down fashion. Instead of the simple block matching metric, CPM uses SIFT-Flow features [109] to compute costs, since this provides more robust matches. SIFT-Flow is a pixel-dense formulation of SIFT [55] and returns a pixel-dense *descriptor field* for a given image. It can be computed more efficiently than the sparse descriptor since the regular, dense setting makes it possible to share intermediate binning results among overlapping descriptors.

In this work, we found that the runtime of CPM can be decreased even further by switching from costly SIFT descriptors to binary descriptors. Such binary descriptors can be computed and matched more efficiently, and they have a smaller memory footprint. However, most binary descriptors with equivalent (or better) ROC performance than SIFT have been obtained with machine learning techniques such as [78,112–114,116] and often exhibit irregular computation patterns, making it difficult to share intermediate results among overlapping descriptors in a dense scan setting. We therefore introduce a new binary descriptor termed BOOM, which has a similar, regular binning cell layout and ROC performance as SIFT – while being computationally more efficient. The proposed descriptor is explained below, followed by a summary of our modifications to the original CPM algorithm. Performance results and comparisons are presented later in Section 3.5.

3.4.1 Binarized Octal Orientation Maps

When looking at existing descriptors, we observe that gradient-based features consistently provide better performance than intensity-based

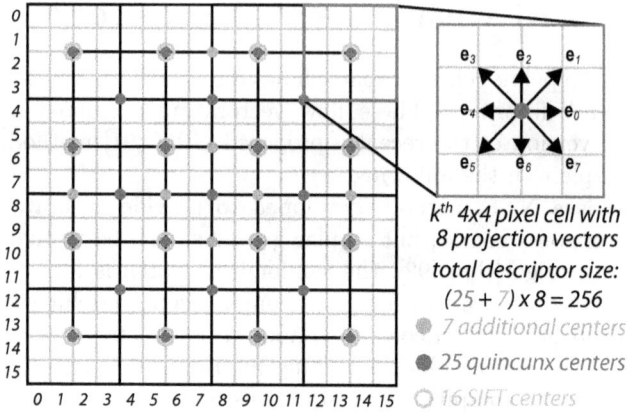

FIGURE 3.8: BOOM descriptor layout: the descriptor is applied to a 16 × 16 pixel patch and contains 32 4 × 4 pixel bins with 8 orientation map responses. This results in 256 values overall. The 4×4 spatial binning cells are arranged in a similar manner as in SIFT, but instead of using 16 non-overlapping cells, we use an arrangement of 32 partially overlapping ones (25 quincunx centers plus 7 additional centers).

ones [112–114]. Well performing descriptors (binary and non-binary) such as BinBoost [112], SIFT [55], DAISY [90] or SURF [56] therefore employ HOG [111] or variations thereof as basic building blocks. Also, normalization of the descriptor vector is essential for good ROC performance [55, 78, 112]. The proposed BOOM descriptor combines HOG-like features with a normalized binarization scheme in order to get a descriptor with similar performance and regular layout as SIFT and DAISY, but with the efficiency of binary descriptors.

Support Region and Orientation Maps

The input to our BOOM descriptor is a normalized 18×18 pixel patch $\mathbf{P}$, which we first preprocess with Prewitt operators in order to get the gradients $\mathbf{g}_j = [G_j^h, G_j^v]$ within a 16×16 pixel support $\mathcal{S}^{16 \times 16}$:

$$\mathbf{G}^h = \mathbf{P} * \begin{bmatrix} 1 & 0 & -1 \\ 1 & 0 & -1 \\ 1 & 0 & -1 \end{bmatrix}, \mathbf{G}^v = \mathbf{P} * \begin{bmatrix} 1 & 1 & 1 \\ 0 & 0 & 0 \\ -1 & -1 & -1 \end{bmatrix}. \quad (3.15)$$

BOOM is formulated as a collection of orientation maps [90] on 4×4 pixel cells (spatial bins) arranged in a quincunx pattern, as shown in Figure 3.8. We additionally add 7 bins around the center region to align the amount of bins with a power of 2. The orientation maps are then built within these 32 spatial bins using eight directions as

$$b_{ki} = \sum_{j \in \mathcal{N}_k^{4 \times 4}} \max \left(0, \langle \mathbf{e}_i, \mathbf{g}_j \rangle \right), \tag{3.16}$$

where b_{ki} are the orientation map responses and $\mathcal{N}_k^{4 \times 4}$ denotes the 4×4 pixel neighbourhood of the kth spatial bin. Overall this results in $32 \cdot 8 = 256$ orientation map responses b_{ki}. Note that we use vector projections onto the following direction vectors

$$\left[\mathbf{e}_0 \ldots \mathbf{e}_7 \right] = \begin{bmatrix} 1 & 1 & 0 & -1 & -1 & -1 & 0 & 1 \\ 0 & 1 & 1 & 1 & 0 & -1 & -1 & -1 \end{bmatrix} \tag{3.17}$$

in order to calculate the gradient contribution to a specific orientation map. Negative projections are clamped to zero. Compared to standard HOG binning approaches, this scheme has the advantage that angles do not have to be calculated explicitly. Note, that the use of unnormalized vectors and Prewitt masks is intentional here as this allows to implement these operations with integer additions only.

Normalization and Binarization

It is crucial to normalize these bin responses in order to get good descriptor performance [78,112]. However, straightforward normalization requires the computation of costly $\mathcal{L}_2$ norms and divisions. As we aim to binarize the final responses anyway, we use a slightly different approach. First, we compute a simple approximation (from [27], page 89) of the average gradient magnitude which can be carried out with integer arithmetic as

$$s = \sum_{j \in \mathcal{S}^{16 \times 16}} 5 \cdot \max \left(|g_j^0|, |g_j^1| \right) + 3 \cdot \left(|g_j^0| + |g_j^1| \right) . \tag{3.18}$$

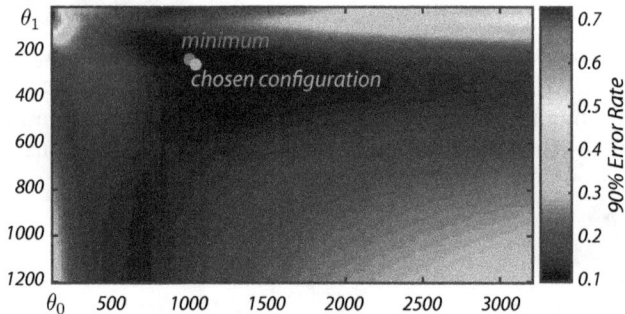

FIGURE 3.9: Sweep of the parameters θ_0 and θ_1 on the UBC Liberty 100 k dataset. The 90 % error rate is color coded and below 10 % for the chosen parametrization.

The normalization of this approximation with $(5+3) = 8$ is implicitly carried out in the binarization step

$$d_{ki} = \begin{cases} b_{ki} \cdot \theta_0 > s & \text{, if } i \text{ even,} \\ b_{ki} \cdot \theta_1 > s & \text{, if } i \text{ odd,} \end{cases} \tag{3.19}$$

which yields a binary 256 bit descriptor d_{ki}. The parameters θ_0 and θ_1 compensate for all normalizations omitted so far. Figure 3.9 shows the resulting 90% ROC error rates. As illustrated in Figure 3.9, we swept these parameters over the range of interest on the Liberty 100 k dataset of Brown et al. [78] in order to determine the optimal values. The minimal error is achieved for $\theta_0 = 980$ and $\theta_1 = 230$. We align these values to powers of two (1024 and 256), since this can be implemented with simple bitshifts and negligible performance impact.

The BOOM descriptor can be efficiently implemented using only integer arithmetic. No divisions, trigonometric- and transcendental functions are required. As elaborated in more detail in Section 3.5, its ROC performance is similar to SIFT descriptors with 128 entries and to DAISY descriptors with 200 entries, while at the same time being more efficient than state-of-the-art binary descriptors in terms of processor execution time. Similar to SIFT-flow and DAISY, BOOM can be implemented as an efficient dense scan method.

Difference to Binarized HOG Descriptors

Although the proposed descriptors has similarities with BHOG descriptors that have been proposed in the domain of human detection [117,118] and sketch-based image retrieval [119], the following important differences should be noted:

1. Usually, HOGs descriptors employ non-overlapping cell layouts, whereas our descriptor uses partially overlapping cells arranged in a quincunx pattern.

2. BHOG descriptors employ gradient binning – i.e., accumulation of the gradient magnitude into corresponding direction bins – instead of orientation maps.

3. BHOG descriptors use either fixed thresholds or the average bin response within a single 4×4 pixel cell to threshold the bin responses. In contrast, our BOOM descriptor uses the average gradient magnitude over the complete descriptor support.

As shown in Figure 3.10 in Section 3.5, these design decisions have a significant impact on the descriptor performance. In addition, the proposed binarization technique does not rely on the final HOG binning values. The binarization threshold can hence be accumulated in parallel to the orientation map responses, and does not require the evaluation of high-dimensional vector norms.

3.4.2 Modifications of CPM and Parametrization

Our CPM closely follows the implementation of [91], and therefore we refrain from repeating all details of the method here and only mention important differences:

- Instead of SIFT-Flow, we use a dense scan implementation of BOOM. In addition, we applied CLAHE [121] with threshold $\theta_{clahe} = 1$ to the input images in order to improve performance on low-contrast image regions.
- Instead of performing the forward-backward check on the two finest levels, we perform one check on the coarsest and one check on the finest level. Outliers on the coarsest level are re-initialized. In addition, we also threshold the matching costs

in order to remove very bad matches. Currently, this threshold is set to $\theta_{desc} = 88$.

- We use the same amount of pyramid levels ($n_{levels} = 5$), propagation iterations ($n_{prop} = 6$), and a grid spacing of $d = 3$. However, we employ a slightly relaxed search radius r of 11 instead of 4, and reduce the forward-backward check threshold θ_{flow} from 3 to 1 pixel as we found that this improves the accuracy of the method.

- The original CPM method does provide sub-pixel accurate results. In order to improve the precision for small flow-vectors, we perform a quadratic interpolation step by reusing the matching costs on 3×3 neighborhoods around matches returned by CPM.

3.5 Results and Comparisons

First, we provide implementation details, followed by an evaluation of BOOM and its combination with CPM. Then, we evaluate the optical flow obtained with our STEA pipeline, and we show results for additional image-based applications.

3.5.1 Implementation and Choice of Parameters

We provide an efficient single core implementation of the complete filtering pipeline in plain C++. Apart from the _popcnt64 SSE intrinsic for computing the descriptor matching cost, we do not make use of other SSE/AVX vector intrinsics in order to facilitate comparisons. The descriptor calculation and CPM parts only use integer arithmetic (with the exception of the subpixel interpolation). The filtering stages are implemented using single precision arithmetic. Timings for the different substeps of our pipeline have been measured on an i7-5557U machine (3.1 GHz) with 16 GB RAM. Detailed numerical results are given in the following subsections.

Note, that there is significant acceleration potential by leveraging vector intrinsics, multiple threads or GPU implementations – especially for high resolution video content where data-level parallelism becomes abundant (Nehab et al. [122] provide useful insights of how recursive filtering approaches can be parallelized). The only part of the current single-core implementation that cannot be trivially parallelized is the sequential search and propagation loop of CPM. For parallel implementations, this issue can be resolved by switching to more elaborate PatchMatch variants such as, e.g., those based on the *jump flood scheme* [92, 123] which use search patterns that are amenable to parallelization.

The filter parameters used to compute the results presented are listed in Table 3.1, and have been tuned on a few frames of the MPI Sintel training set [93]. Parameters for spreading sparse data have been validated on the MPI Sintel test set by means of the optical flow evaluation in Section 3.5.3, and work well for a variety of different sequences and applications. We set the fidelity term λ to 0.0 since it is not intended to be used with sparse data and can lead to discontinuities at the sparse sampling locations. For applications with artistic intent (such as, e.g. *Visual Saliency* and *Base/Detail Layer*, see Section 3.5.4), the selected filter parameters represent a possible choice and may be changed in order to smooth the images at different scales. The employed parameterization of the temporal filter provides a good tradeoff between oversmoothing and no smoothing at all, as explained in more detail in Section 3.5.4. For CPM, we use the parameters as described in Section 3.4.2. The only varying parameter is the grid spacing d which we set according to the image resolution such that CPM yields approximately the same amount of flow vectors (~30-40 k). For 'Sintel' images (1024×436) we use $d = 3$ and for 'Tears of Steel' (TOS) images (1920×800) we use $d = 6$.

TABLE 3.1: Filter parameters employed.

Parameter	Flow	Disparity	Saliency	Base/Detail
σ^{XY}	0.017	0.017	0.025	0.025 / -
σ^{photo}	0.3	0.3	0.3	0.3
σ^{grad}	1.0	1.0	1.0	1.0
n_{iter}	5	5	5	5 / 0
$\lambda^{\{XY,T\}}$	0	0	1	1
$\alpha^{\{XY,grad,photo\}}$	2	2	2	2

3.5.2   Performance of CPM and BOOM

BOOM Descriptor

We first evaluate the BOOM descriptor on the three 100 k datasets *Liberty*, *Notredame* and *Yosemite* provided by Brown et al. [78], where each contains 100 k pairs of rotation- and scale-normalized 64×64 grayscale image patches. The pairs come with annotated ground truth, which allows to calculate performance measures in terms of ROC[4]. This dataset allows for a precise comparison of different descriptors as there is no need for interest point detection and descriptor support extraction. Note, that the scale parameter needs to be adjusted for each descriptor method for fair comparison [78].

As mentioned in Section 3.4.1, the design decisions to use orientation maps instead of HOGs, and to calculate the normalization threshold on the complete descriptor support have a significant impact on the descriptor performance. Figure 3.10 shows an ROC performance comparison of the proposed BOOM descriptor and three variants thereof that either employ standard HOG binning with 8 directions instead of the orientation maps, or that use simple per 4×4 pixel cell normalization as described in [118]. Note the performance increase due to orientation maps and complete support normalization.

Figure 3.11 shows the ROC of our BOOM descriptor, along with several other standard descriptor methods [44, 55–60]. With a 95 % error rate of 22-27.3 %, we can observe that it performs slightly better than SIFT (23.3-29.5 %) and DAISY (22.7-31.7 %), and it performs significantly better than many other standard binary descriptors such as BRIEF, BRISK, FREAK and SKB. The similar performance of BOOM, SIFT and DAISY can be attributed to the fact that all work on similar gradient bin arrangements. The novelty of BOOM lies in the combination of orientation maps with a normalized thresholding scheme, that allows to retain this performance in spite of binarization.

---

4. The ROC captures the TPR to FPR tradeoff and can be used to measure the accuracy of a descriptor when used to match image interest points. The true positive rate is defined as TPR = #TP/(#TP + #FN) and the false positive rate is defined as FPR = #FP/(#FP + #TN), where #TP, #FP, #TN, #FN denote the numbers of true/false positives/negatives, respectively. It is desirable to have a high TPR and a low FPR.

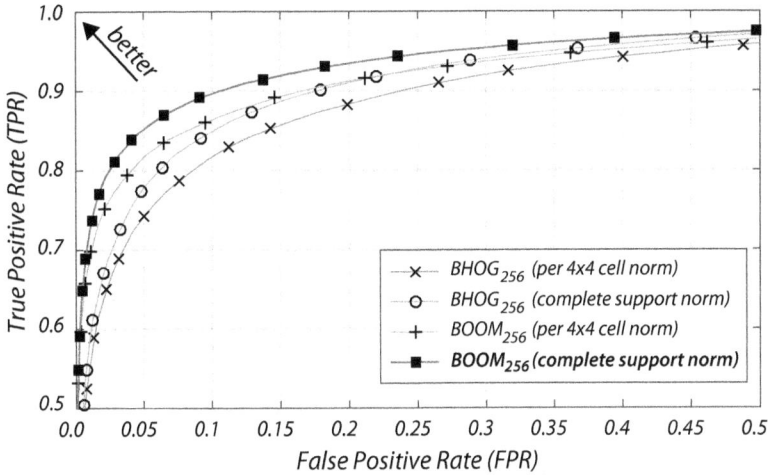

FIGURE 3.10: Comparison of the ROC performance (Notredame 100k UBC dataset [78]) of the proposed BOOM descriptor with 3 variants with the same spatial bin layout, but that selectively use standard HOG binning instead of orientation maps, or standard per cell HOG normalization instead of the proposed complete support normalization.

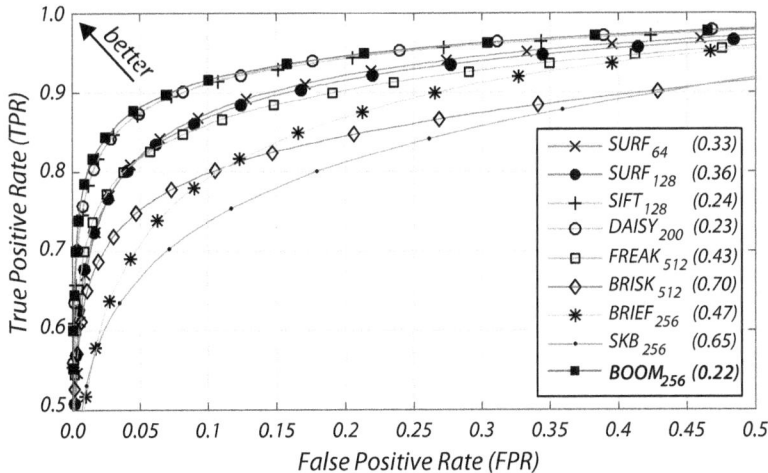

FIGURE 3.11: ROC comparison of BOOM with other standard methods on the Notredame 100 k UBC dataset [78].

TABLE 3.2: BOOM performance and comparison with other state-of-the-art descriptors on the UBC dataset [78]. The 95 %error rates are for the 100 k datasets if not stated otherwise. Timings are for single descriptors and have been obtained by extracting 10 k descriptors from the same image including descriptor support normalization.

Descriptor	Type	Size [B]	95% Error+ Liber.	95% Error+ Notred.	95% Error+ Yosem.	Time [µs]	Implementation	Processor Model
SIFT$_{128}$ [55]	FP	128	29.54	23.65	29.27	483.6	VLFeat [124]	i7 3.1 GHz (1 Thread)
SURF$_{128}$ [56]		128	44.72	36.13	42.77	69.8	MATLAB	i7 3.1 GHz (1 Thread)
SURF$_{64}$ [56]		64	43.28	32.53	42.16	63.1	MATLAB	i7 3.1 GHz (1 Thread)
Brown$_{29}$ [78]		29	**17.56**	**11.98**	**13.55**	-$^\times$	-	-
DAISY$_{200}$ [90]		200	28.33	22.66	31.71	-$^\times$	author	-
BinBoost$_{64}${128} [112]	Bin	**8**	21.08†	15.72†	20.93†	~1000§	author	Mobile i7 2.66 GHz
LDA-Hash$_{128}$ [116]		16	49.66	51.58	52.95	-	author	-
Bamboo$_{128}$ [113]		16	49*	-	-	~20§	author	-
LDDB-U$_{256}$ [114]		32	-	-	-	~154§	author	ARM Cortex A9 1 GHz
FREAK$_{512}$ [59]		64	52.10	43.48	46.63	34.4	MATLAB	i7 3.1 GHz (1 Thread)
BRISK$_{512}$ [58]		64	73.74	70.24	67.55	38.2	MATLAB	i7 3.1 GHz (1 Thread)
BRIEF$_{256}$ [125]		32	53.47	46.51	51.12	29.9	author	i7 3.1 GHz (1 Thread)
SKB$_{256}$ (Chapter 2)		32	67.07	64.95	62.42	**10.9**	ours	i7 3.1 GHz (1 Thread)
BOOM$_{256}$		32	27.27	22.04	27.23	**10.9**	ours	i7 3.1 GHz (1 Thread)

* using the 200k dataset. + 95% Error is the FPR for a TPR=0.95. § time taken from original publication. † runs using different training sets have been averaged [112]. × not available since DAISY is implemented as a dense scan method.

More detailed numerical results are provided in Table 3.2, together with results from advanced learning-based descriptors such as Brown et al. [78], LDA-Hash [116] and AdaBoost-based binary descriptors like BinBoost [112] and its simplified derivatives Bamboo [113] and LDDB-U [114]. We observe that in general, binary descriptors have much lower memory footprints with respect to the FP counterparts (assuming 1 B per dimension for FP methods [112]). Especially BinBoost achieves a very compact representation with just eight bytes while ranking second in terms of 95 % error rates. However, it is very costly to compute (~1 ms per descriptor) as each effective descriptor dimension results in the computation of a linear superposition of 128 features. Therefore, this method is better suited for large-scale database retrieval where memory and matching efficiency matter most. As opposed to this, BOOM offers a competitive speed/performance tradeoff, since with an execution time of ~11 μs per descriptor, it is significantly more efficient than most other methods. At the same time it provides similar error rates as SIFT and DAISY.

The run-time of the dense scan DAISY and SIFT-Flow methods (published MATLAB MEX C++ implementations) amounts to 1350 ms and 950 ms, respectively, for a 1024 × 436 pixel frame on a 3.1 GHz i7 machine. Compared to this, our unvectorized single-core dense scan implementation of BOOM only requires 111 ms, which is 12.3× faster than DAISY and 8.6× faster than SIFT-Flow.

The SKB descriptor does not perform well in this setting as it has been designed for strict stereo setups. This behavior can be attributed to the SKB filter kernels which are quite sensitive to distortions in the image. SKB is therefore not suited for large differences in camera perspective that are present in the UBC dataset and which can occur in optical flow applications. Note however, that SKB outperforms most other descriptors (Chapter 2) in strict stereo setups.

CPM with BOOM

Figure 3.12 shows a comparison of the sparse flow estimates delivered by the original CPM method (calculated using the code from [91] with standard parameters) and our CPM variant with BOOM. The corresponding numerical results using the metrics defined in [91] are shown in Table 3.3. SIFT-NN are matches obtained with SIFT and

Input Frames   **Ground Truth**

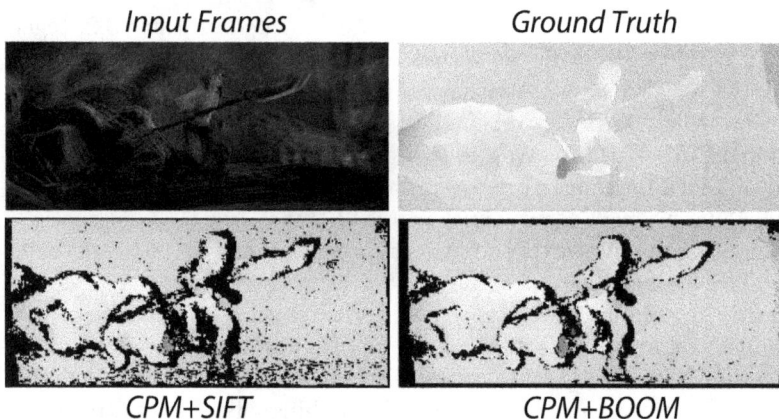

CPM+SIFT   *CPM+BOOM*

FIGURE 3.12: Comparison of CPM+BOOM with CPM+SIFT from [91]. Images copyright ©, BLENDER FOUNDATION, www.sintel.org.

FLANN [91], *Kd-tree PatchMatch* is a recent NNF method [126], *Deep Matching* is the matching method used in *DeepFlow* [107] and *EpicFlow* [106], and CPM+SIFT is the original CPM method [91]. We can see that CPM+BOOM performs similarly as CPM+SIFT in terms of density and precision. At the same time, it is more than twice as fast when run on a slower i7 machine (3.1 GHz) than the original (3.5 GHz). The decreased number of matches of CPM+BOOM when compared to CPM+SIFT is mainly due to the larger border region of our implementation and a stricter forward-backward consistency check. This decrease can be tolerated since the method provides sufficiently dense initializations.

TABLE 3.3: Comparison of CPM+BOOM with related methods on the MPI Sintel final training set. The execution times of CPM+SIFT and CPM+BOOM have been measured on the same processor model, whereas the remaining timings are from [91].

Method	Amount	Density	Precision	Time	Processor
SIFT-NN [55]	1K	0.175	0.851	0.5 s	i7 3.5GHz
KPM [126]	**446K**	**1.000**	0.595	**0.4 s**	i7 3.5GHz
Deep Matching [107]	5K	0.892	0.945	15 s	i7 3.5GHz
CPM+SIFT [91]	40K	0.886	0.975	2.1 s	i7 3.1GHz
CPM+BOOM	32.7K	0.834	**0.979**	0.56 s	i7 3.1GHz

3.5.3 Optical Flow Performance

Evaluation

We evaluated the flow estimation part of our pipeline on the MPI Sintel dataset [93], that is considered realistic and challenging, and which exhibits large motion and motion blur. The set comprises 23 training and 12 test sequences with up to 49 frames. Being derived from an animation movie, accurate ground-truth is available for training.

Table 3.4 lists our evaluation results, together with published results of the fastest methods (run times below 20 s per frame). A visual comparison of selected methods is shown in Figure 3.13, and the run time vs. AEE tradeoff is shown in Figure 3.14. As can be observed in Figure 3.13, our method provides visually similar quality as CPM-Flow [91], EpicFlow [106], and FlowFields [127] which however provide more accurate results in terms of AEE. The lowest AEE is achieved by CPM-Flow on the clean pass (3.56) and by FlowFields on the *final pass* (5.81). However, the run time of our unvectorized single-core implementation amounts to only 0.59 s for XYT filtered flow results and is therefore significantly faster than most other methods listed in Table 3.4. This fact is also reflected in the plot in Figure 3.14, which summarizes the quality/speed tradeoff of all methods compared. We can see that our BOOM +PF provides a very competitive tradeoff, and shares the Pareto frontier with PCA-Flow, CPM-Flow and FlowFields which are optimal choices at different operating points. The only CPU-based method that is faster than BOOM +PF is PCA-Flow [108], which only requires 0.18 s. But as can be seen in Figure 3.13, PCA-Flow is not capable of capturing fine details and the resulting flow is blurry[5].

It is interesting to note that all methods on the Pareto frontier (PCA-Flow, CPM-Flow and FlowFields) rely on similar concepts to calculate the flow. They extract sparse point correspondences or

---

5. A remark on the AEE metric is in order here. Although widely accepted as quality metric for optical flow methods, it should be used with care and the results should also be inspected and compared visually. The AEE is a simple global measure and errors such as small missing details, or blurry boundaries are not always captured correctly. Also, the actual value of the AEE difference from one method to another can be misleading. E.g., the PCA-Flow and PCA-Layers methods are very close to our method in terms of AEE– yet they visually look significantly worse.

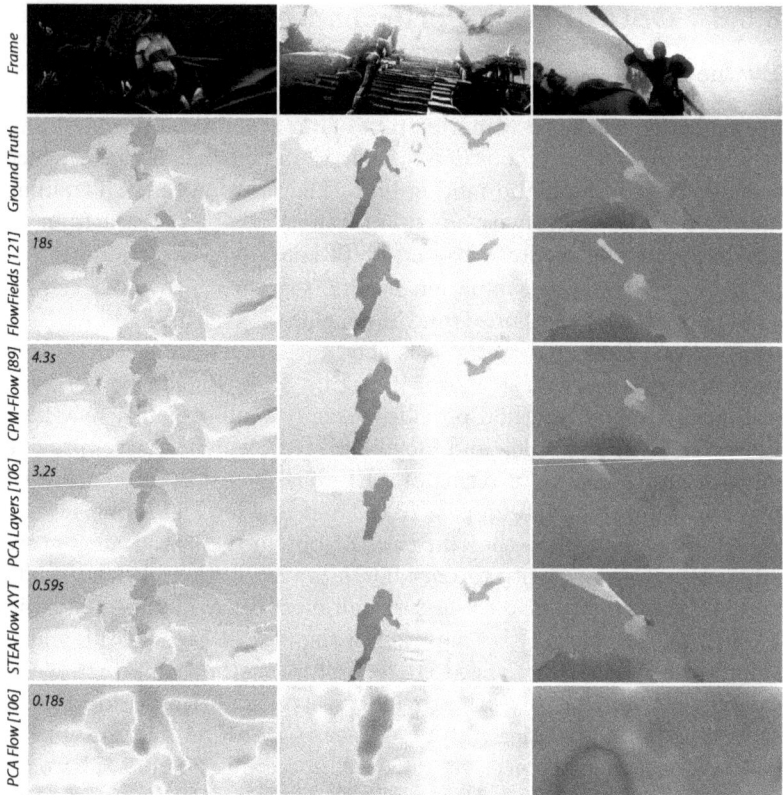

FIGURE 3.13: Flow results and comparison of our method on the MPI Sintel dataset. Images copyright ©, BLENDER FOUNDATION, www.sintel.org.

NNFs, and convert these to a dense optical flow via interpolation. The superior accuracy of CPM-Flow and FlowFields is achieved by additional optimization of the interpolated flow using the variational refinement backend of EpicFlow. Such a refinement could also be added to our method, but this would lead to longer runtimes.

TABLE 3.4: Results and comparison with other fast state-of-the-art methods on the MPI Sintel dataset [93]. Timings taken from [91].

Method	Clean Pass AEE			Final Pass AEE			Time [s]	Processor Model
	All	Noc	Occ	All	Noc	Occ		
TF+OFM [128]	4.92	1.87	29.7	6.73	3.39	33.9	~400 s	i7 3.5GHz (1 Thread)
DiscreteFlow [129]	3.57	1.11	23.6	6.08	2.94	31.7	~180 s	i7 3.5GHz (1 Thread)
SparseFlow [105]	6.20	2.36	37.5	7.85	3.86	40.4	10 s	?
DeepFlow [107]	5.38	1.77	34.8	7.21	3.34	38.8	19 s	i7 3.5GHz (1 Thread)
FlowFields [127]	3.75	**1.06**	25.7	**5.81**	**2.62**	31.8	18 s	i7 3.5GHz (1 Thread)
EpicFlow [106]	4.12	1.36	26.6	6.29	3.06	32.6	16.4 s	i7 3.5GHz (1 Thread)
CPM-Flow [91]	**3.56**	1.19	**22.9**	5.96	2.99	**30.2**	4.3 s	i7 3.5GHz (1 Thread)
PCA-Layers [108]	5.73	2.46	32.5	7.89	4.26	37.5	3.2 s	i7 3.5GHz (1 Thread)
BOOM+PF (XY)	5.20	1.70	33.8	7.60	3.73	39.1	0.54 s	i7 3.1GHz (1 Thread)
BOOM+PF (XYT)	5.31	1.82	33.8	7.68	3.83	39.1	0.59 s	i7 3.1GHz (1 Thread)
PCA-Flow [108]	6.83	3.01	37.9	8.65	4.73	40.7	**0.18 s**	?

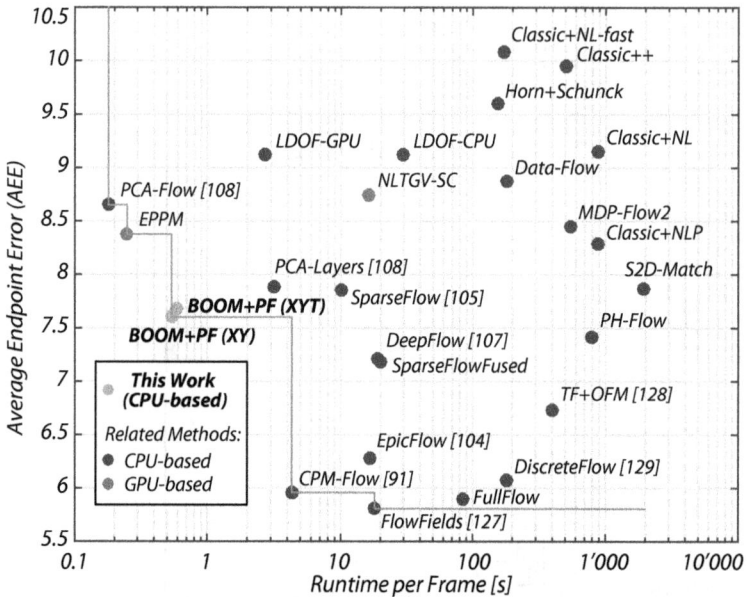

FIGURE 3.14: Runtime vs. AEE plot of fast CPU implementations on the *final* MPI Sintel dataset. The Pareto front is indicated in gray.

The small numerical AEE difference between the XY and the XYT filtered flows is due to the fact that an asymmetric, causal filtering window is used. On one hand, this can lead to a small temporal lag which is reflected in a slightly increased AEE metric. On the other hand, the T step effectively removes temporal high-frequency noise as shown in Section 3.5.4.

Performance and Related STEA Methods

Calculating optical flow for a Sintel frame (1024×436) takes 0.59 s (without file I/O), and most of the time is spent calculating the dense descriptor field (26 %) and performing CPM (50 %). The five spatial filtering iterations of the two flow components and the confidence map amount to 16 %, and the temporal step including warping to only 8 % of the total runtime. For this resolution, CPM requires a large fraction of the time. Note however, that the computational effort of CPM does not increase when scaling to higher resolutions,

TABLE 3.5: Comparison of our STEA pipeline with Lang et al.'s method on a 640×480 sqeuence with N frames. 'Frame' is abbreviated with 'f' in this table.

Method	Iter. XY/T	Time [s/f]	Lat. [f]	Mem. [MB]	BW [MB/f]	Processor Model
Lang et al. [86]	4/4	0.63	N	N·22	?	i7 2.67 GHz (4 Th.)
This work	5/1	0.35	1	50	322	i7 3.1 GHz (1 Th.)

since we keep the amount of sparse grid sampling points constant. For larger frames such as in the TOS examples (1920×800) shown in the following subsections, the fraction spent in CPM reduces to 25 %. The runtime for one frame amounts to 1.41 s in that case, and the remaining fractions are 39 % for BOOM, 24 % for the XY iterations, and 11 % for the T step. The DRAM occupancy of our method does not exceed 72 MB for Sintel and 245 MB for TOS frames. The DRAM traffic per frame has been profiled by observing last-level data cache misses, and amounts to ~500 MB and ~1.66 GB, respectively.[6]

The related STEA method by Aydın et al. [81] on which our spatial filter is based, has not been designed to calculate flow, and relies on a slow, high-quality variational method [104]. The STEA method by Lang et al. [86] is able to calculate optical flow, but has not been thoroughly evaluated on a large dataset like Sintel. As summarized in Table 3.5, their optimized quad-core C++ implementation requires 0.63 s on average to process a 640×480 pixel image including feature matching and four XYT iterations. Our single-core implementation requires only 0.35 s for the same image including CPM+BOOM, 5 XY iterations and one T step. Note, that the required amount of memory and the processing latency of the method by Lang et al. grows linearly with the number of frames, whereas our method has a constant latency and memory footprint. E.g., their method requires 8.83 GB of DRAM to process 400 frames of 640×480 video, while our method works with less than 50 MB of DRAM for any sequence with that resolution.

---

6. Typical embedded GPU platforms such as the NVIDIA Tegra X1 and X2 offer in the order of 25-50 GB/s of DRAM bandwidth [20].

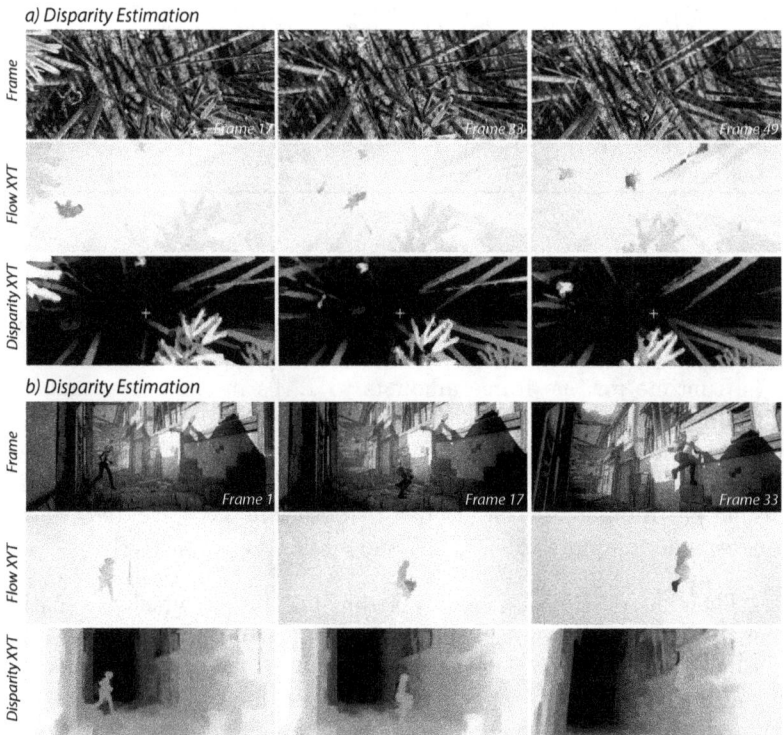

FIGURE 3.15: Disparity estimation results from our pipeline. Images copyright ©,
BLENDER FOUNDATION, www.sintel.org.

3.5.4   Applications and Temporal Consistency

In addition to optical flow, our STEA pipeline can be used for a
variety of important image-based applications for which we present a
selection of qualitative results in this section[7].

Disparity Estimation

Similarly to flow estimation, disparity estimation involves computations of dense correspondences between image pairs. The main
difference is the geometrically constrained stereoscopic setup, and the
fact that both images are from the same time instant. This allows
to additionally apply RANSAC to filter correspondence outliers [28].

Our STEA method can be used to produce temporally consistent disparity maps by initializing the additional channels **A** with sparse point correspondences, which are then first propagated and aligned with the input image, and then temporally smoothed in the T step. In Figure 3.15a and b we show two example sequences with video footage from the MPI Sintel stereo dataset. The sparse disparity initialization has been obtained with the same CPM+BOOM matching routine used for calculating optical flow vectors.

Base/Detail Layer Decomposition

Base/detail layer decompositions[8] obtained with edge-aware filters are powerful tools enabling HDR tone mapping methods [81, 83, 96] and detail manipulation methods [82, 96, 101]. In Figure 3.16a, we show one instance of such a method, where details of a video sequence are enhanced. Using our STEA filter, we compute the base layer by $\mathbf{B}_t = \mathbf{I}_t^{XYT}$, where $\mathbf{I}_t$ are the input video frames. A detail layer is then computed by subtracting this base layer from only temporally filtered input frames as $\mathbf{D}_t = \mathbf{I}_t^T - \mathbf{B}_t$. As described in [96], the detail layer is now boosted by multiplying each pixel by a factor ($3\times$ in the example shown) and applying a properly shifted and normalized sigmoid curve in order to avoid hard clipping. The enhanced image is then obtained by adding the boosted detail layer back to the base layer.

Visual Saliency

A saliency map identifies the visually important regions in the image [34], and is an important feature for many image-based applications such as retargeting and multiview rendering based in IDW [14, 32]. Efficient methods like [35] analyze the frequency spectrum of the image and since they operate on a per-frame basis, they often produce temporally noisy output. Our STEA pipeline can be used to stabilize and clean such noisy data, as show in Figure 3.16b, where the per-frame saliency Guo et al. [35] has been filtered with our pipeline.

---

7. See also http://iis.ee.ethz.ch/~michscha/stea/ for videos of the examples.

a) Base/Detail Layer Decomposition

b) Visual Saliency

FIGURE 3.16: a shows examples of base-detail layer decompositions, and b shows examples of visual saliency estimation. Our pipeline has been used remove temporal flickering. Images copyright © BLENDER FOUNDATION, mango.blender.org.

8. Such decompositions are useful tools for manipulating image details or colours. They can be thought of as filter banks that decompose an image into several layers containing different spatial frequency bands. These layers can be manipulated separately and are again combined at the end to produce the enhanced image. See [81–83, 96, 101] for more details.

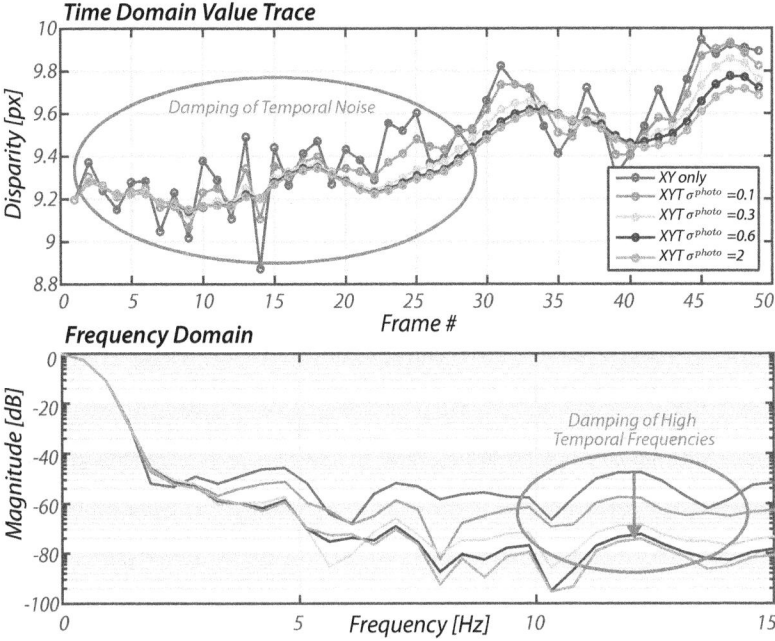

FIGURE 3.17: Impact of the temporal filter: a shows a disparity value trace at the position of the green cross in Figure 3.15a, and b shows the corresponding behaviour in the frequency domain.

Temporal Consistency

We demonstrate the impact of the T filter by means of a value trace taken at the position with a green cross in the disparity example in Figure 3.15a. This value trace is shown in Figure 3.17a for different temporal σ^{photo} parameterizations. We observe that the T filter effectively removes high-frequency flickering noise for parameter values in the order of 0.1-0.3. This behaviour is also reflected in the corresponding frequency domain analysis in Figure 3.17b. We used $\sigma^{photo} = 0.3$ to compute all presented results since this provides a good tradeoff between oversmoothing and flickering removal. For the parameterization of the gradient measure, we found $\sigma^{grad} = 1.0$ to be a well-performing choice for the applications at hand.

3.6   Summary and Conclusions

We have designed a STEA filtering pipeline which can be used to approximate challenging regularization problems such as optical flow estimation, or to temporally stabilize noisy image-based feature maps such as visual saliency. Below we list the key take-away's:

- Our pipeline is more efficient than previous methods [81, 86] since the incremental PF formulation does not require large temporal neighbourhoods to be available in memory. Also, our pipeline does not depend on precomputed optical flow as, e.g., [81], since it is able to bootstrap its own optical flow on-the-fly by interpolating a quasi-dense NNF obtained with an improved patch match method. With a runtime of only 0.59 s per frame and an AEE of 7.683 px on the MPI Sintel dataset our method provides a competitive speed/quality tradeoff when compared to several other state-of-the-art optical flow methods.

- For accurate and fast optical flow initialization, we improve an existing PatchMatch NNF method called CPM method [91] by switching from SIFT-Flow-based features to a new binary descriptor. The descriptor termed BOOM has been designed specifically for this task, and outperforms other binary state-of-the-art descriptors in terms of ROC performance. As opposed to BOOM, the binary SKB descriptor from Chapter 2 does not perform well in this setting as it has been designed for rectified stereo setups. SKB is very sensitive to differences in camera perspective an therefore not suited for optical flow applications where camera motion can be quite large. Note however, that SKB outperforms most binary descriptors in strict stereo setups (see also Appendix 3.A for a comparison).

- Similar to existing STEA approaches [86], our method trades accuracy for computational efficiency, and is therefore not without limitations. The quality of the obtained results depends on the amount and quality of the input images and initial conditions. In case of sparse flow vectors, the proposed CPM+BOOM matching method alleviates this problem to a certain degree, since it provides accurate, quasi-dense initializations. However, similar to other matching methods it is dependent on the input image quality and the presence of noise and blur may lead to false correspondences. Our method can also fail when important object boundaries are not well represented by image edges – a limitation which applies to many other image-based applications and EA filters. Despite these limitations, the presented pipeline offers an efficient way to apply EA filtering to video streams, without the requirement for large temporal windows. Apart from fast optical flow estimation, the pipeline can be used to create temporally consistent results for a variety of important image-based applications such as disparity and saliency estimation, and methods working on base/detail layer decompositions.

- Compared to state-of-the-art methods, our STEA pipeline has lower computational cost and constant memory requirements, making it better suitable for embedded GPU or hardware implementations. E.g., it could be used to enhance temporal consistency of real-time video processing systems such as video retargeting [130] and MVS engines (Chapter 7).

The algorithmic improvements presented in this chapter significantly reduce the computational complexity and memory requirements of STEA filtering. A natural next step is to demonstrate real-time operation by implementation on an ASIC/FPGA or GPU target. In this context, there are a couple of opportunities for future work:

- The BOOM features involved in the NNF computation can be efficiently calculated and are well parallelizable. However, the randomized search step in CPM is more challenging to implement on a parallel architecture. There exist solutions for SW-based implementations such as the *jump-flood scheme* [92, 123], but such schemes are not straight-forward to implement in HW. So there is an opportunity to investigate efficient parallel implementations and HW architectures, as NNFs are important features for many image and video processing applications [92].

- EA filters such as the PF and DT operate iteratively on vertical and horizontal scan lines, and the external DRAM bandwidth linearly increases with the number of applied iterations. This is

because the filtered channels and auxiliary arrays do not entirely
fit into the data caches anymore if full-HD footage is processed.
A preliminary implementation showed that *tiled* filtering approaches can be effective at recovering data locality and have
the potential to significantly reduce the external bandwidth. In
addition, they may increase the available data-level parallelism.
Therefore, such tiling could be key for efficient parallel implementations. It should however be noted, that results can be nu-
merically different, and additional techniques such as deblocking
filters have to be used to reduce tiling artifacts.

- Other applications can benefit from the techniques developed in
  this chapter. The NNF calculated with CPM and BOOM is very
  robust and achieves significantly better image coverage than
  standard, interest-point-detector-based approaches (Chapter 2).
  Applications that work on sparse depth or flow data are likely
  to benefit from it (see also Appendix 3.C for a short comparison
  in the scope of MVS). For IDW applications in general, the
  STEA pipeline could either be used to filter and stabilize the
  input features, or it could even be used as a replacement for the
  warp calculation step involving the solution of large regularized
  LS problems. STEA filtering scales better to higher resolutions
  than solvers for LS problems (Chapter 4), but at the same time,
  STEA could also introduce artifacts due to the EA filtering
  limitations discussed before.

- Currently, the temporal filter recursion just uses forward warping to align the past filter state with the current frame. This
  works well under the assumption that the quantities warped are
  constant over time. While this assumption usually holds for
  slowly moving content, it may fail in areas with fast moving
  objects. In the current implementation, the temporal gradient
  measure is used to prevent warping and filtering artifacts in
  such areas. However, a better solution may be found by also
  incorporating first and second-order derivatives of the filtered
  quantities into the warping step. Theoretical work along these
  lines may lead to concepts similar to Kalman filtering in linear
  systems theory.

3.A   Comparison of BOOM and SKB

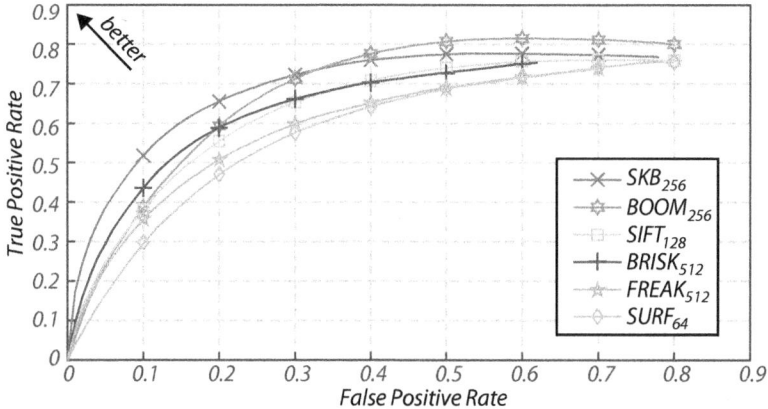

FIGURE 3.18: ROC performance comparison with a greedy nearest-neighbor matching with a small window as in the hardware system presented in Chapter 2. The test set comprises rectified stereo images and is from [131].

We saw in Section 3.5 that BOOM outperforms many other state-of-the-art descriptors when matching interest points from image pairs with large differences in viewing position. We also observed that SKB performs not very well in such settings. Therefore, we repeat a similar evaluation as in Chapter 2 on rectified stereo data [131] in order to assess the performance of BOOM and to provide a more holistic comparison of SKB and BOOM. Interest points in corresponding stereo pairs have been extracted using the DOB detector, and these interest points are then matched using either SKB, BOOM, SIFT, BRISK, FREAK or SURF. The resulting ROC curves are shown in Figure 3.18. We can see that SKB clearly outperforms all other descriptors (even BOOM) in the low-threshold region on the left side. We can therefore conclude that neither of the two descriptors (SKB and BOOM) is better than the other in all cases. The two methods rather complement each other, and while SKB delivers high recognition rates for controlled, rectified setups, BOOM is more robust and delivers better results when the camera viewing positions undergo larger changes of perspective.

3.B A BOOM Feature Extraction Core

In order to compare the hardware complexity of BOOM with other descriptors, we adapted the SKB ASIC from Chapter 2. In particular, we exchanged the SKB descriptor calculation part with a unit that calculates the BOOM descriptor. The interest point detector has been left unchanged. Preliminary synthesis of the resulting chip are given in Table 3.6, together with synthesis results of the SKB core in 180 nm and 65 nm. A floorplan of the BOOM core can be found in the chip gallery (Section A.11). Both the BOOM and the SKB cores have been designed for similar throughput in terms of fps and descriptors per frame. We can observe that the descriptor calculation units for BOOM and SKB have almost identical complexity in terms of kGE. This is to some extent expected, since both have almost equal runtimes in SW (see Table 3.2). BOOM requires somewhat more intermediate memory due to the fact that the binarization threshold is only available once all intermediate results have been calculated. The higher overall chip complexity of the BOOM chip is due to a much larger interest point FIFO between the interest point detection and descriptor calculation stages (depth 128 instead of 20). This helps to significantly reduce the amount of dropped interest points in image regions with high density.

TABLE 3.6: Comparison of the SKB and BOOM feature extraction cores (synthesis).

Resources	SKB Core[§]	SKB Core[§]	BOOM Core
Technology	180 nm	65 nm	65 nm
Conditions	TT 25 °C	SS 125 °C	SS 125 °C
Core Voltage	1.8 V	1.2 V	1.2 V
Descriptor Logic	75 kGE	73 kGE	70 kGE
SRAM	29 kbit (80 kGE)	29 kbit (85 kGE)	33 kbit (103 kGE)
Total Complexity	223 kGE	224 kGE	262 kGE
Performance			
Frequency	100 MHz	300 MHz	300 MHz
Resolution	720p	1080p	1080p
Throughput	30 fps	40 fps	40 fps
IP Fifo Depth	20	20	128
Max. Desc./Frame	12 k-25 k	27 k-56 k	25 k-40 k

§ Partially improved version of the SKB core (leaner FIFOs, improved SKB kernels from Appendix 2.A) that is used in the MVS system (Chapter 7).

3.C MVS with Nearest Neighbour Fields

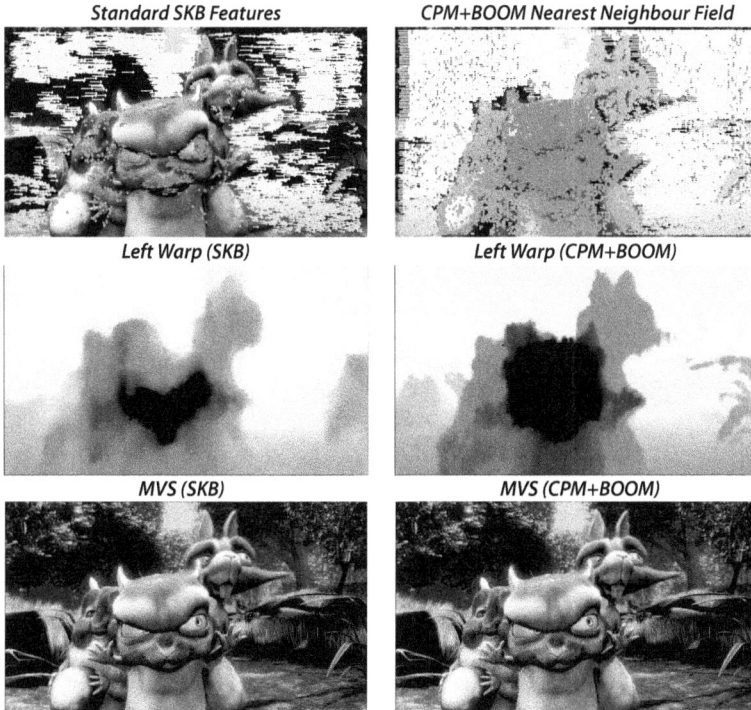

FIGURE 3.19: Comparison of MVS using standard, interest-point-detector-based SKB features (Chapter 2), and a sparse NNF computed with CPM+BOOM. The NNF has significantly better image coverage, low-outlier rate, and the resulting 180×100 warp captures more details. The MVS results are in anaglyph format.

A preliminary evaluation shows that MVS can significantly benefit from the sparse NNF features (CPM+BOOM) that have been developed in this chapter. As can be seen in Figure 3.19, the NNF has significantly better image coverage than the standard approach. Therefore, small details are better captured, and the resulting warp better reflects the scene geometry. Note however, that an NNF is more costly to compute than standard point correspondences, since several random search iterations are performed on the complete image, whereas the standard features can be computed in one scanline pass (with some local buffering, see Chapter 2).

Chapter 4

Evaluation of Linear Solvers for IDW

4.1 Introduction

Several computational photography- and computer vision applications have in common that they involve an optimization step where certain quantities (such as camera-/object positions and pixel values/-coordinates) are optimally estimated given multiple measurements such as images or extracted image features. Popular application examples include simultaneous localization and mapping (SLAM) [132], HDR compression, optical flow estimation and inpainting [133, 134], or IDW applications such as video retargeting, disparity remapping and MVS [14, 29, 32]. Such optimization problems are often posed as LS problems, which essentially boil down to solving large, sparse systems of linear equations. Solving these problems in real-time within the power budget of mobile SoCs is challenging. This motivates the development of custom hardware accelerators able to solve common subtasks such as LS problems with a minimum of energy.

In this chapter, we present an architectural exploration of direct and iterative linear solvers for LS problems emanating from IDW applications such as [14, 29, 32]. In particular, the aim is to compare the energy efficiency of solver variants that make use of off-chip

DRAM memory, and solvers that use on-chip SRAM exclusively. This is a very important design aspect since the use of a DRAM based solution has system-level implications in terms of modularity (the overall bandwidth has to be shared among all subsystems in an SoC), energy efficiency and additional cost such as memory controller IP, DRAM components and I/Os. In this light, it may be preferable to accept a somewhat larger datapath in return for being able to store all data in on-chip memories, i.e., for doing away with any external DRAM. A basic estimation framework has been developed to compare these architecture variants in terms of energy dissipation, circuit area, throughput and precision. We highlight potential tradeoffs to guide architectural and system-level decisions, and show that in 28 nm CMOS technology, complete on-chip solvers are perfectly feasible for matrix dimensions up to $128\,k^1$.

The remainder of the chapter is organized as follows: Sections 4.2 and 4.3 summarize related work and some preliminaries. The evaluated hardware architectures are explained in Section 4.4, our evaluation framework is explained in Section 4.5, and experimental results are presented in Section 4.6. Finally, Section 4.7 summarises the main takeaways of the chapter.

4.2 Related Work

The most common iterative and direct solvers used in LS problems are conjugate gradients (CG) and Cholesky decomposition (CHD) solvers, since they exploit the mathematical properties of such problems in order to achieve high efficiency [135–138]. The design of hardware architectures for these iterative and direct solver types has been addressed in [33, 139] and [43, 140–143], respectively. In all of the listed work, FPGA implementations are considered, and most use relatively small, non-sparse matrices – except Greisen et al. [33] and Schaffner et al. [43]. In the former, direct CHD and iterative CG solver architectures for large, sparse, linear systems in image and video processing applications are developed and implemented on an FPGA. In the latter it is shown that in the case of IDW, an

---

1. This is sufficient for high quality IDW of full-HD video in most cases, as evaluated in Section 7.14.

incomplete CHD can provide an approximate solution that is still accurate enough – yet with significantly lower computational cost. These two references serve as a basis for the hardware architectures evaluated in this chapter.

The estimation framework used for our exploration is based on analytical architecture and activity models, and a set of pre-characterized submodules (such as FP operators). An example for a similar framework is the one by Vishnoi et al. [144] used to analyze different configurations of QR accelerators. Our designs, in contrast, are significantly larger and also comprise SRAM macros and DRAM interfaces, which are a key part of the design and must be accounted for. In this work, estimates for the external DRAM components are obtained from the MICRON System Power Calculator [145]. Other popular estimation frameworks such as CACTI [146] and McPAT [147] are not suitable for our application since they focus on general purpose cache- and multi-processor architectures.

4.3 Preliminaries

In this section we first give a short introduction into the problem structure arising from IDW applications before summarizing the target solver algorithms evaluated in detail in the following sections.

4.3.1 Linear Systems in Image/Video Processing

Many image and video processing algorithms can be posed as a quadratic minimization problem of the form

$$\min_{\mathbf{f}} \left(E\left(\mathbf{f}\right) \right) = \min_{\mathbf{f}} \left(E_d\left(\mathbf{f}\right) + E_s\left(\mathbf{f}\right) \right), \tag{4.1}$$

where the data term E_d enforces function values at certain sampling positions, and the smoothness term E_s is a regularizer that propagates the known sample values to adjacent sampling positions. The vector $\mathbf{f}$ is holding the samples of an unknown discrete function (e.g., pixel intensities or coordinate values) defined on a two-dimensional sampling grid Λ of width w and height h. The two terms E_d and E_s are briefly summarized below, a more detailed description can be found in Chapter 7. Let i be the *linear* index of the sampling points of the grid Λ with $i \in \mathcal{D}_\Lambda = \{1, 2, ..., w, ..., w \cdot h\}$. Figure 4.1 a) illustrates the relations introduced for one sample f_i of $\mathbf{f}$. The data term usually has the form

$$E_d\left(\mathbf{f}\right) = \sum_{i \in \mathcal{D}_\Lambda} \lambda_i \left(f_i - p_i\right)^2, \qquad (4.2)$$

where p_i are the constraint values, and the parameters λ_i are weights indicating the relative importance of the corresponding constraints. The smoothness term contains differential constraints defined among neighboring samples, and is usually given by

$$E_s\left(\mathbf{f}\right) = \sum_{i \in \mathcal{D}_\Lambda} \left(\lambda_i^x \left(f_{i+1} - f_i - d_i^x\right)^2 + \lambda_i^y \left(f_{i+w} - f_i - d_i^y\right)^2\right), \quad (4.3)$$

for a 5-stencil (Figure 4.1a)), where d_i^x and d_i^y are the constraint values, and λ_i^x and λ_i^y are again relative-importance-weights. The superscripts ^{x} and ^{y} indicate whether the parameter belongs to a horizontal or vertical difference constraint.

The constraints can be chosen in such a way that the quadratic energy functional (4.1) can be written in matrix form as

$$\min_{\mathbf{f}} \|\mathbf{A}\mathbf{f} - \mathbf{b}\|^2, \qquad (4.4)$$

where $\mathbf{A} \in \mathbb{R}^{N \times M}$, $\mathbf{f} \in \mathbb{R}^M$ and $\mathbf{b} \in \mathbb{R}^N$. In this case, the energy minimization problem is equivalent to an LS problem, and can be solved very efficiently. N is the amount of overall constraints, and $M = w \cdot h$ the amount of variables (grid vertices) to be solved for. The smoothness constraints (4.3) act as a regularizer and ensure that $N > M$, i.e., $\mathbf{A}$ is a 'tall' matrix leading to an overdetermined LS problem. The solution to this problem can be found by solving the associated normal equations

$$\mathbf{A}^T\mathbf{A}\mathbf{f} = \mathbf{A}^T\mathbf{b}, \qquad (4.5)$$

where $\mathbf{A}^T\mathbf{A}$ is an $M \times M$ matrix. Since the constraints are defined on small, local stencil neighborhoods on Λ as shown in Figure 4.1a), the matrix $\mathbf{A}^T\mathbf{A}$ is very sparse and only contains a main- and a few off-diagonals (Figure 4.1b). Further, the number of variables in the linear system is in the order of tens of thousands to millions – depending on the resolution of Λ.

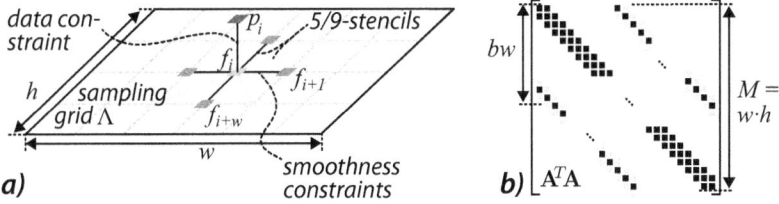

FIGURE 4.1: a) Sampling grid with unknowns f_i and constraints on a 5-stencil/9-stencil. b) Sparsity structure of the resulting normal equations matrix $\mathbf{A}^T\mathbf{A}$. Due to the symmetry, only 5 (off-) diagonals are unique.

4.3.2 Matrix Properties in IDW Applications

The IDW applications considered in this work non-linearly transform the frames of a video in *content-adaptive* manner. Video retargeting is concerned with changing the aspect-ratio of the video frame, e.g., from 16:9 to 4:3. Automatic MVS uses IDW to inter- and extrapolate a given S3D frame-pair to new virtual view positions. The constraints for the energy minimization problems are cues extracted from the original video frames, and the solution vector $\mathbf{f}$ represents the coordinates of the pixels in the transformed image. Key for those applications is a so called *saliency* map [34] which is used to weigh smoothness constraints in (4.3). This effectively moves distortions to visually unimportant regions in the image. The energy minimization problems usually also include:

1. edge- or line constraints,
2. a few position constraints on the image border,
3. disparity constraints in the case of S3D video,
4. and temporal constraints.

Such temporal constraints are of the form

$$E_d = \sum_{i \in \mathcal{D}_\Lambda} \lambda_i \left(f_i[t] - f_i[t-1] \right)^2, \qquad (4.6)$$

where $\mathbf{f}[t-1]$ is the solution computed from the previous time step. This constraint is important since on one hand it ensures that there are no temporal 'wobbling' artifacts in the transformed frames. On the other hand, it improves the condition of the problem matrix significantly [43]. This also holds true for other data constraints such as point correspondences in MVS. The resolution of current video content is predominantly 1080p (1920 pixels wide by 1080 pixels in height), resulting in two (one for each coordinate dimension) to four (when S3D footage is used) equation systems with nearly two million variables for each frame in the video. Solving such large systems at frame rates of up to 30 fps is computationally very demanding. Therefore, the problems are usually solved on grids sub-sampled by a factor of ten or so to reduce the computational complexity. This results in realistic grid sizes of about 180×100, which equals to ~18 k variables in the minimization problem. For the IDW applications at hand, the problem matrix $\mathbf{A}^T\mathbf{A}$ usually has one main-diagonal and four[2] off-diagonals – out of which two are unique due to the symmetry (Figure 4.1 b). Another property of the normal equations is that $\mathbf{A}^T\mathbf{A}$ is square and positive definite, which is important for choosing an appropriate solver.

4.3.3 Linear Solver Algorithms

Linear solver algorithms fall into two main categories, namely *direct* and *iterative* ones. Direct solvers employ a matrix decomposition (such as LU, LL, QR or SVD decompositions) in order to compute an exact solution in one shot, whereas iterative solvers successively refine an approximate solution estimate. The choice of a particular solver is dependent on several factors such as the mathematical properties and structure of $\mathbf{A}^T\mathbf{A}$, convergence and numerical behaviour, and the complexity of the arithmetic operations and memory accesses involved. To find the subset of suitable solvers for the LS problems at hand, we can follow the guidelines by Barret et al. [137] and Demmel et al. [138] and first consider that there are two main approaches how the LS problem in (4.4) can be tackled:

1) Solution of the system of normal equations $\mathbf{A}^T\mathbf{A}\mathbf{f} = \mathbf{A}^T\mathbf{b}$. This can – in principle – be achieved by using any linear solver, but the fact that $\mathbf{A}^T\mathbf{A}$ is symmetric, positive definite allows to narrow down the selection of solvers. The most efficient solvers in terms of computational complexity for this type of matrices are direct CHD solvers or iterative CG solvers[3].

---

2. Note, that when using subsampled grids, the data constraints need to be interpolated, resulting in four additional off-diagonals right next to the outermost off-diagonals of $\mathbf{A}^T\mathbf{A}$ (Chapter 7).

2) Minimization of the expression (4.4). This can be achieved using either the SVD decomposition, or by using specialized iterative solvers for LS problems such as, e.g., CGLS [148] or LSQR [149]. These iterative solvers do not explicitly work with the normal equations, but rather with the matrix $\mathbf{A}$ and its transpose $\mathbf{A}^T$. Although these methods produce the same iteration sequence as CG solvers in exact arithmetic, they exhibit better numerical properties in case of ill-conditioned matrices and finite precision. However, this usually comes at the cost of higher complexity (memory and computations) within each solver iteration.

The first approach is conceptually simpler, as it not only allows to use fast and lean algorithms for symmetric and positive definite matrices, but also the amount of normal equations remains constant. I.e., while the size of $\mathbf{A}$ and $\mathbf{b}$ varies with the number of data constraints N, the size of $\mathbf{A}^T\mathbf{A}$ is always constant. As shown by Greisen et al. [150], it is possible to derive analytical formulations for the elements of $\mathbf{A}^T\mathbf{A}$. These can be evaluated on-the-fly without explicitly building $\mathbf{A}$ and $\mathbf{b}$. The second approach provides better numerical properties in general, since matrix $\mathbf{A}$ is typically better conditioned than the product $\mathbf{A}^T\mathbf{A}$[4]. However, the algorithms are more complex and the size of $\mathbf{A}$ and $\mathbf{b}$ is variable.

In the following, we will adopt the first approach, as it is conceptually simpler and therefore preferable for a hardware implementation. In addition, the IDW problems at hand are typically well conditioned for the matrix dimensions considered as will be explained in Section 4.3.5.

---

3. A side note on the biconjugate gradient stabilized (BiCGStab) solver implemented by Greisen et al. [150] is in order here. Although BiCGStab can be used to solve the normal equations, it is not the most efficient solver for this problem (according to [137]). In fact, it has been developed for non-symmetric matrices, can exhibit irregular convergence behaviour, and requires about twice as many working vectors and linear algebra operations per iteration. A comparison between preconditioned conjugate gradients (PCG) and BiCGStab on IDW problems showed that BiCGStab does not converge faster than PCG, and therefore it has not been considered further in this evaluation.

4. This is because $\operatorname{cond}\left(\mathbf{A}^T\mathbf{A}\right) \approx \operatorname{cond}\left(\mathbf{A}\right)^2$, see also [136, 137].

4.3.4   Direct Cholesky Solver

The CHD comes in two variants, namely the $\mathbf{LL}^T$ and the $\mathbf{LDL}^T$ decompositions[5]. In both cases, the $\mathbf{L}$ matrix is a lower-triangular matrix that can be used to obtain a solution to the equation system $\mathbf{A}^T\mathbf{A}\mathbf{f} = \mathbf{A}^T\mathbf{b}$ with a forward and a backward substitution step. The difference is that in the second variant, the diagonal elements of $\mathbf{L}$ are offloaded to the diagonal matrix $\mathbf{D}$. We only consider the second variant here, since it is more amenable to hardware implementations [143] due to fewer divisions and the absence of square-roots.

The computational burden and storage requirements of a CHD based solver grow with $\mathcal{O}\left(M^3\right)$ and $\mathcal{O}\left(M^2\right)$, respectively, and are in general much higher than for an iteration of CG. However, the sparsity structure of $\mathbf{A}^T\mathbf{A}$ leads to a banded Cholesky factor $\mathbf{L}$ with bandwidth $bw = h + 1$ or $bw = h + 2$ (depending on the stencil), where h is the height of the image grid. The complexities then reduce to $\mathcal{O}\left(M^2\right)$ and $\mathcal{O}\left(M^{1.5}\right)$.

4.3.5   Incomplete Cholesky Factorization

A so called *incomplete Cholesky factorization* can also be used in order reduce the complexity of the decomposition step. The idea is to drop insignificant elements in the Cholesky factor $\mathbf{L}$ below a certain threshold ρ during the evaluation of large scalar products between the rows of $\mathbf{L}$. While this scheme is often used to calculate pre-conditioners for iterative methods [136], we realize that incomplete factorizations are precise enough to directly compute a solution in the case of the IDW applications at hand. This becomes evident by looking at Figure 4.2a) and b), where we can observe that the average magnitude of the 'fill-in' elements of $\mathbf{L}$ can be very small compared to the main diagonal elements – especially for IDW problems where

---

5. Fill-reducing permutations such as *nested dissection* [151] and *approximate minimum degree ordering* [152] have not been considered further in this evaluation, since they destroy the regular band structure of the matrix. A preliminary HW implementation indicates that, while reordering is a viable choice for systems with *one single* processing element, it is difficult to extract fine-grained parallelism required for real-time operation. In addition, reordering impedes scan-line processing of the in- and outputs, and complicates the architecture considerably due to the use of sparse data formats and irregular memory access patterns.

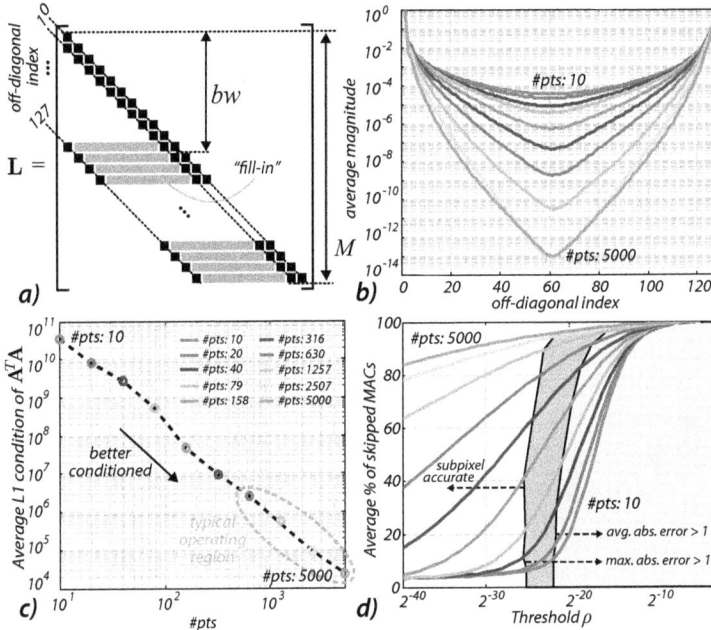

FIGURE 4.2: Evaluation of incomplete Cholesky factorizations on a set of ten MVS problems with 32 k variables and with a varying amount of random data constraints (10 to 5'000 per frame). a) shows the banded structure of **L**. b) shows the average magnitude of the elements of **L**, averaged along the diagonals. c) shows the averaged L1 condition estimate (condest from MATLAB) of the problem matrices. Finally, d) shows the percentage of MAC operations that can be skipped as a function of the threshold ρ. The regions where the maximum- and mean errors are greater than one pixel are shaded in gray. The results shown have been obtained using double precision, but the technique can also be combined with lower-precision arithmetic.

many data constraints (#pts) are available. The L1 condition of the matrix $\mathbf{A}^T\mathbf{A}$ strongly correlates with the amount of data constraints as can be seen in Figure 4.2c). Due to the temporal data constraint and other point features used to formulate data constraints, the typical operating region is at the right side of this plot, where matrices are well conditioned. As show in Figure 4.2d), many 'fill-in' elements can be dropped with only small impact on the final solution accuracy, and the MAC operations dependent on these elements can be skipped, leading to a large reduction of MAC operations (over 60% in this example). However, this comes at the cost of a variable runtime dependent on the matrix condition. For more details on this evaluation, see [43].

4.3.6   Iterative CG Solver

Iterative solvers such as CG start from an initial solution $\mathbf{f}_0$ which is then successively refined according to a specific recurrence equation (shown in Algorithm 1 for PCG). An advantage of iterative solvers is that they can be easily parallelized, and that their memory and computational complexity is in the order of $\mathcal{O}(M)$ for one iteration, where M is the dimension of the matrix $\mathbf{A}^T\mathbf{A}$. However, they have to be properly preconditioned in order to ensure rapid convergence, and require a high memory bandwidth due to their iterative nature. In the CG variant employed, a standard diagonal pre-conditioner $\mathbf{m} = \text{diag}\left(\mathbf{A}^T\mathbf{A}\right)$ is sufficient, since the $\mathbf{A}^T\mathbf{A}$ matrices are typically well conditioned in our application.

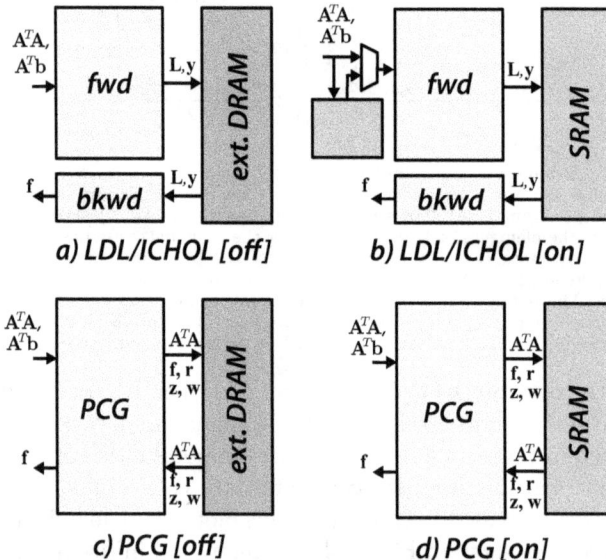

FIGURE 4.3: On- and off-chip memory configurations of the solver templates considered. a) and b) are direct solvers (either an $\mathbf{LDL}^T$ decomposition or an incomplete Cholesky decomposition), and c) and d) are iterative, preconditioned CG solvers. The direct solvers decompose the matrix $\mathbf{A}^T\mathbf{A}$ into $\mathbf{L}$ and an intermediate solution $\mathbf{y}$ in the *fwd* path, and compute the final solution $\mathbf{f}$ in the *bkwd* path. The CG solvers iteratively refine an approximate solution $\mathbf{f}$.

4.4 Evaluated Hardware Architectures

In this exploration, we consider parametrizable architecture templates of the following three solvers:

- **LDL** is a standard CHD solver that factorizes $\mathbf{A}^T\mathbf{A}$ into a lower triangular matrix $\mathbf{L}$ and a diagonal matrix $\mathbf{D}$.
- **ICHOL** is the same except that an incomplete factorization is carried out by dropping elements of $\mathbf{L}$ that have a magnitude lower than a certain threshold ρ.
- **PCG** denotes a CG solver with Jacobi preconditioner.

Each architecture template comes in an on-chip and an off-chip memory configuration, that either stores intermediate vectors and decomposition data in on-chip SRAM or off-chip DRAM, respectively. These variants are denoted with the additional labels [on] and [off], as illustrated in Figure 4.3 a-d). All architecture templates have some additional parameters such as the memory size or the parallelization degree, which will be explained in more detail below.

4.4.1 Cholesky Variants

The architecture of the standard CHD solver is shown in Figure 4.4, and consists of two parts: The first part performs the decomposition and the forward-substitution (these can be interleaved in the same datapath), and the the second part carries out the backward substitution. Although the backward pass is similar to the forward pass, matrix $\mathbf{L}$ and vector $\mathbf{y}$ have to be accessed in reversed order (bottom-up), and therefore this task can only be executed once the decomposition and forward pass have been finished. Using a separate unit is more convenient and allows to calculate both the forward- and backward pass of two subsequent matrices in parallel.

The decomposition stage contains a scalar product of n_{par} elements which is used to calculate the inter-row products in column-major order. The decomposition is sequential in nature, since each element L_{ij} depends on all it's neighbours to the left. However, due to the banded shape of $\mathbf{L}$, the values required to compute another column of $\mathbf{L}$ all lie within a window of size bw^2. These are buffered locally, together with the past bw elements of the $\mathbf{y}$ vector and $\mathbf{D}$

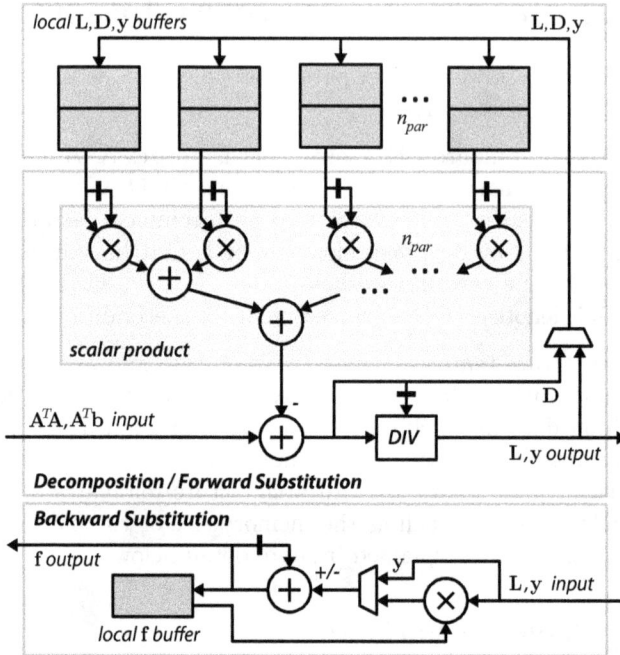

FIGURE 4.4: Datapath template of the LDL solver. The forward path calculates **L** and the forward substitution in interleaved manner. The backward path operates in parallel and executes the backward substitution of a previously computed decomposition.

matrix. Parallelization is easy up to a degree of $n_{par} = bw$. Beyond that, the strong dependencies on previous results impedes further parallelization. Here, the parallelization degree is specified relative to the matrix bandwidth and indicated as postfix in the architecture name. E.g., LDL[off][$bw/2$] corresponds to $n_{par} = \lceil bw/2 \rceil$.

As mentioned before, two variants of LDL with different memory configurations are considered in this evaluation. The first, LDL[off] stores **L** and **y** in the off-chip DRAM (Figure 4.3a). The second variant LDL[on] uses only on-chip SRAM to store **L** and **y**. Since **L** can be very large ($bw \cdot h \cdot w$ entries) it may not be feasible to store the whole matrix on chip. Therefore we also consider variations of LDL[on], where only a fraction of **L** is kept in the memory (the rest is discarded). The solution **f** is then computed with several

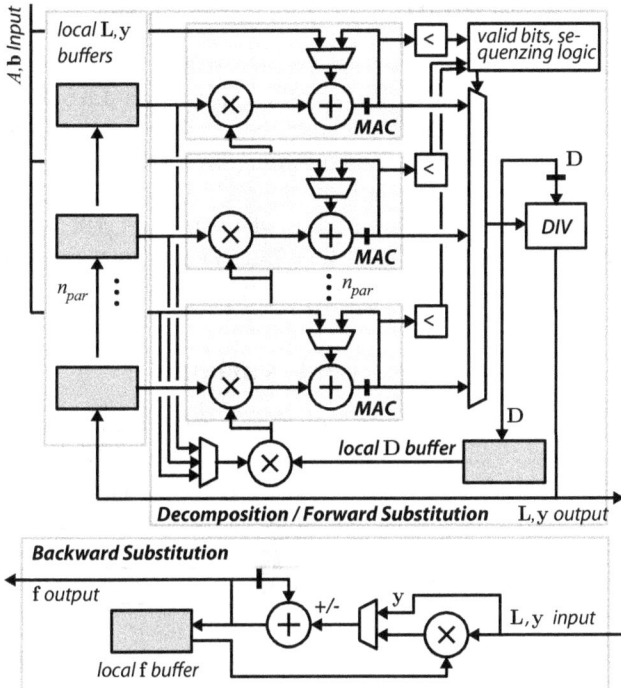

FIGURE 4.5: Datapath template of the ICHOL solver. The forward path contains additional logic to perform the thresholding, while the backward path is the same as in the LDL solver in Figure 4.4.

decomposition passes. Consider for example an on-chip buffer that can store half of **L**. Overall, two decomposition passes are then needed: first, the whole decomposition is calculated, but only the bottom half of **L** is stored. The bottom half can then be used to perform the backward substitution, while the upper half of the decomposition is recomputed and stored. This variation requires less on-chip memory, but needs more time due to the re-computation steps. Therefore, for all LDL[on] variants, the parallelization degree n_{par} is set to bw, and the parameter in the postfix denotes the fraction of memory required to store the complete **L** matrix on-chip. E.g., LDL[on][$m_D/2$] is able to store half of **L** on-chip.

4.4.2   Approximate Cholesky Variants

The ICHOL variants are similar to LDL except that they take advantage of the zeroes in $\mathbf{L}$ as described in Section 4.3.5. As a consequence the scalar products cannot be computed in parallel using a rigid multiplier-adder tree structure. Instead, these scalar products are computed sequentially in parallel by individual MAC units (Figure 4.5). But the computed $\mathbf{L}$ elements are still gathered in column-major order. In the evaluation, the memory configurations are the same as for LDL (Figures 4.3a and b).

4.4.3   Conjugate Gradient Variants

The PCG solver iteratively works on the residual vector $\mathbf{r}_k = \mathbf{A}^T\mathbf{A}\mathbf{f}_k - \mathbf{A}^T\mathbf{b}$ (k is the iteration index) and produces several intermediate vectors $\mathbf{r}_k, \mathbf{p}_k, \mathbf{w}_k, \mathbf{f}_k, \mathbf{z}_k$, as shown in Algorithm 1. The initial solution $\mathbf{f}_0$ can be arbitrary and is chosen to be the all-ones vector here. The

---

Algorithm 1 Jacobi Preconditioned CG.

1: **procedure** $\mathrm{PCG}(\mathbf{A}^T\mathbf{A}, \mathbf{A}^T\mathbf{b}, \mathbf{m}, \epsilon^2, n_{maxIter})$
2:      $\beta_0 \leftarrow 0, t_0 \leftarrow 1, k \leftarrow 1$ ▷ Initialize
3:      $\mathbf{r}_0 \leftarrow \mathbf{A}^T\mathbf{b} - \mathbf{A}^T\mathbf{A}\mathbf{f}$
4:      $\mathbf{z}_0 \leftarrow \mathbf{r}_0./\mathbf{m}$ ▷ Preconditioner solve
5:      $b^2 \leftarrow \left(\mathbf{A}^T\mathbf{b}\right)^T \mathbf{A}^T\mathbf{b} = \mathbf{b}^T\mathbf{A}\mathbf{A}^T\mathbf{b}$
6:      **while** $\mathbf{r}_k^T\mathbf{r}_k/b^2 > \epsilon^2$ **and** $k < n_{maxIter}$ **do**
7:          $\mathbf{p}_k \leftarrow \mathbf{z}_{k-1} + \beta_{k-1}\mathbf{p}_{k-1}$ ▷ Pass 1:
8:          $\mathbf{w}_k \leftarrow \mathbf{A}^T\mathbf{A}\mathbf{p}_k$
9:          $\alpha_k \leftarrow t_{k-1}/\left(\mathbf{w}_k^T\mathbf{p}_k\right)$
10:         $\mathbf{f}_k \leftarrow \mathbf{f}_{k-1} + \alpha_k\mathbf{p}_k$ ▷ Pass 2:
11:         $\mathbf{r}_k \leftarrow \mathbf{r}_{k-1} - \alpha_k\mathbf{w}_k$
12:         $\mathbf{z}_k \leftarrow \mathbf{r}_k./\mathbf{m}$
13:         $t_k \leftarrow \mathbf{r}_k^T\mathbf{z}_k$
14:         $\beta_k \leftarrow t_k/t_{k-1}$
15:         $k \leftarrow k + 1$
16:     **end while**
17:     **return** $\mathbf{f}_k$ ▷ The solution vector
18: **end procedure**

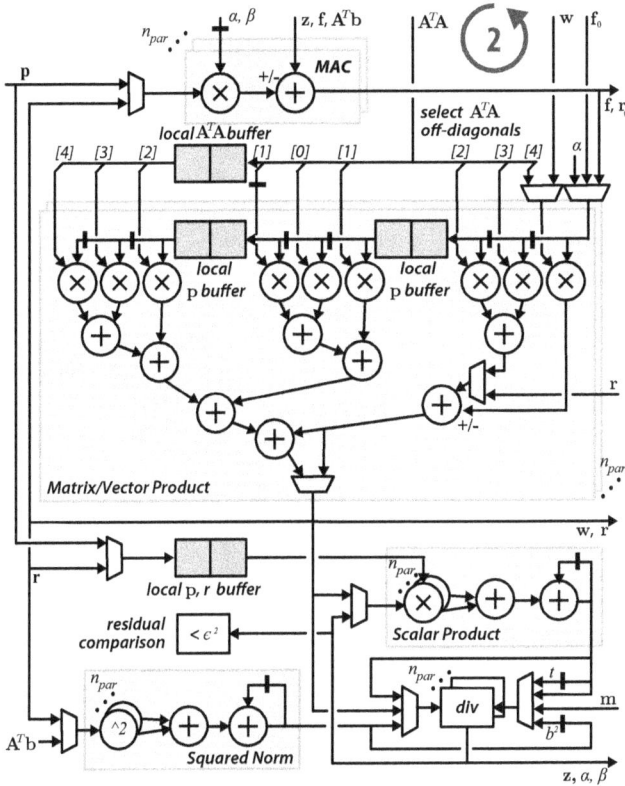

FIGURE 4.6: Datapath template of the PCG variants. One iteration is computed in two passes due to data dependencies in Algorithm 1.

pre-conditioner matrix is $\mathbf{m} = \mathrm{diag}\left(\mathbf{A}^T\mathbf{A}\right)$. The parameter ϵ denotes the relative residual[6] tolerance and is used as the iteration stopping criterion. Its value is determined later in the evaluation.

Both PCG variants evaluated have the same datapath, shown in Figure 4.6. They only differ in the way the intermediate results are stored – i.e., PCG[off] stores everything off-chip whereas PCG[on] stores everything on-chip (see Figures 4.3c and d). The datapath consists of several MAC units for the vector-vector product calculation, several matrix-vector multipliers, and one scalar-product. Since

---

6. The relative residual is defined as $\|\mathbf{r}_k\|/\|\mathbf{A}^T\mathbf{b}\| = \|\mathbf{A}^T\mathbf{A}\mathbf{f}_k - \mathbf{A}^T\mathbf{b}\|/\|\mathbf{A}^T\mathbf{b}\|$.

the matrices at hand contain at most 5 or 9 (off-)diagonals, the matrix-vector multipliers can evaluate one row-vector product per cycle. Due to the sequential dependency on the scalar-products α and β, two passes are required per PCG iteration. However, the vector-vector, matrix-vector, and scalar products can be parallelized easily. The amount of parallelization is determined by the parameter n_{par}, and will be used as a postfix in the architecture identifier. E.g., PCG[off][4×] would correspond to a PCG solver with off-chip memory and fourfold parallelization $n_{par} = 4$.

Overall, storage for 9 vectors of dimension M is needed in the case of a 9 off-diagonal matrix: there are 5 unique diagonals of $\mathbf{A}^T\mathbf{A}$ and 4 temporary vectors $\mathbf{p}_k, \mathbf{f}_k, \mathbf{r}_k$ and $\mathbf{w}_k$ ($\mathbf{w}_k$ and $\mathbf{z}_k$ may share the same memory location). Memory accesses are completely linear, and hence the utilization of the off-chip memory bandwidth is close to optimal.

4.5   Estimation Framework

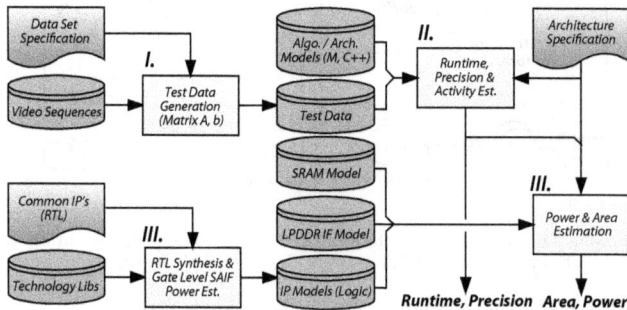

FIGURE 4.7: Flow of the estimation framework used in this work.

In order to quickly assess different architecture variants and find potential tradeoffs a MATLAB-based framework has been developed which estimates *area*, *throughput*, *power consumption* and *precision*. The estimations are based on a combination of analytic architecture models, pre-characterized logic instances, SRAM datasheet values and an LPDDR 3 system power model. It is important to note that the aim is not to achieve very high estimation accuracy, but rather to quickly compare several architecture variants in a simple and efficient manner. In particular, the framework enables early estimations at the system level in order to observe general trends. The estimation flow is shown in Figure 4.7 and contains three main parts:

   I. Data generation
  II. Runtime, precision and activity estimation
 III. Power and area estimation

The three parts are explained in more detail below. Note, that this estimation framework could also be used to explore other types of hardware architectures by adding suitable architecture models.

4.5.1 Test Data Generation

The test data is an integral part of the evaluation since its statistics directly affect the numerical performance of the evaluated architectures. The test data is generated according to parameters specified in a *Data Set Specification*. This step depends on the application – in this case real-world video sequences are analyzed in order to produce the LS problem matrices $\mathbf{A}^T\mathbf{A}$ and the vectors $\mathbf{A}^T\mathbf{b}$. The test data is organized as a batch of sequences which are stored on disk. The architectures to be explored can then be evaluated on these pre-computed batches.

4.5.2 Runtime, Precision & Activity

For each of the hardware architectures to be evaluated, a parameterizable multi-precision software model is implemented in order to estimate the quantization error and other statistics such as matrix non-zero patterns and iteration counts. Together with analytic cost models, these statistics can be used to estimate the throughput of a certain architecture instance and the activities of individual logic and memory elements. The individual parameters of an architecture are defined in an *Architecture Specification*. In fact, such a specification can define several instances, all of which are evaluated on a given test data batch to make parametric explorations.

4.5.3   Power & Area Estimation

Power and area estimations rest on the assumption that the dominant contributions (in both circuit area and power consumption) come from the FP operations, from the on-chip SRAM memories and from the off-chip memory subsystem. These three contributions are modeled separately, as explained below.

Logic

The power dissipation and area of common FP operations such as multiplication, addition/subtraction and division are calculated using SYNOPSYS DesignWare IPs and a gate-level switching activity interchange format (SAIF) back-annotation flow. I.e., the IPs are first wrapped in register transfer level (RTL) code and synthesized with SYNOPSYS Design Compiler. The netlists obtained are then simulated in MENTOR Modelsim and toggle activities are extracted and annotated to a SAIF file, which is then read back into SYNOPSYS Power Analyzer to estimate the static and dynamic power. In this exploration, post layout parasitics are estimated using wireload models provided by the standard cell library vendor. In order to get a worst case estimate, 2'000 random testvectors are used in the simulation. These steps are performed for several clock periods and for different parameterizations of the FP IPs (exponent/mantissa widths). The values obtained are cast into interpolated lookup tables (LUTs) which – together with the runtime and activity values – can then be used to make estimates for arbitrary parameter combinations. For this evaluation, all IPs have been characterized for the typical case corner (0.8 V, 25 °C) of a 28 nm CMOS process.

Internal SRAM

In order to estimate circuit area and power requirements of SRAMs, we extracted the corresponding values from the datasheets of a representative set of high-density, single-port SRAM instances. These values were used to build an interpolated look-up table which allows for making estimates for arbitrarily configured SRAM macros with word widths up to 64 bit, and depths up to 16 k. Since 16 k is the maximum depth that the employed memory compiler can generate, larger SRAMs are assembled from multiple instances.

External DRAM

We use an LPDDR 3 memory interface for this evaluation, which consists of three parts: *memory controller, physical layer (PHY)* and *memory module.* Since the memory architecture is very similar to LPDDR 2, the power and area values for the memory controller were taken from [153], where implementation results for a few controller variants in STM 28 nm technology are presented (the results of the largest memory controller are used). These values are then derated to the final operating frequency and scaled to the correct interface width. An estimate for the PHY area is taken from the same source. However, since no power figures for this PHY are provided, they are estimated using the LPDDR 3 values published by JEDEC [154]. In addition, it is assumed that the data lines are actively terminated to 50 Ω during data transfers (otherwise the termination is switched off). The power dissipation of the external memory module is based on the MICRON System Power Calculator [145] for LPDDR 2. Since the LPDDR 3 spreadsheet is not open-access, we updated it to LPDDR 3 using values from the datasheet of the MICRON EDF8132A1MC LPDDR 3 device. This approach was deemed feasible due to the similarity of LPDDR 2 and LPDDR 3 and since no special power-down modes exclusive to LPDDR 3 are used here. Currently, our model supports the three configurations LPDDR 3-400 with 32 bit, LPDDR 3-800 with 32 bit and LPDDR 3-800 with 64 bit, corresponding to memory bandwidths of 3.2 GB/s, 6.4 GB/s and 12.8 GB/s, respectively.

4.6 Evaluation Results

The operating point for all architectures in this evaluation is chosen to be 400 MHz, since the maximum frequency of the SRAM macros employed lies around 450 MHz. Further, all off-chip variants have a LPDDR 3-800 interface with a maximum bandwidth of 6.4 GB/s, which is a reasonable configuration for mobile devices. For all architectures except the ICHOL variants, an additional logic overhead

TABLE 4.1: Summary of solvers evaluated and their parameterizations. The FP format contains 8 exponent bits, and the #mantissa bits is chosen separately for each grid size. The FP format and ρ and ϵ have been chosen such that subpixel precise solutions with a median error below 0.1 pixel are achieved. The [on] variants store their intermediate results in on-chip SRAM, while the [off] variants use off-chip DRAM.

Direct Solvers	Parallelism n_{par}	Memory Size† [words]	#Mant. Bits / Thresh. ρ		
			64×128	128×256	256×512
LDL[on]	bw	$\frac{m_D}{8}, \frac{m_D}{4}, \frac{m_D}{2}, \frac{m_D}{1}$	28/-	35/-	38/-
LDL[off]	$\frac{bw}{8}, \frac{bw}{4}, \frac{bw}{2}, \frac{bw}{1}$	m_D			
ICHOL[on]	bw	$\frac{m_D}{8}, \frac{m_D}{4}, \frac{m_D}{2}, \frac{m_D}{1}$	$28/2^{-24}$	$32/2^{-27}$	$33/2^{-28}$
ICHOL[off]	$\frac{bw}{8}, \frac{bw}{4}, \frac{bw}{2}, \frac{bw}{1}$	m_D			
Iterative Solvers	Parallel. n_{par}	Memory Size† [words]	#Mant. Bits / Tol. ϵ		
			64×128	128×256	256×512
PCG[on]	4×,8×,12×,16×	m_I	$29/10^{-9}$	$32/10^{-12}$	$36/10^{-13}$
PCG[off]	4×,8×,12×,16×	m_I			

† $m_D = bw \cdot M$ is the memory size needed to store the decomposition data **L** and **y**, and $m_I = 9 \cdot M$ the size needed to store the matrix $\mathbf{A}^T\mathbf{A}$ and the working vectors.

of 10% is assumed. In the case of ICHOL, a higher value of 20% is applied to account for additional logic coming from the two large multiplexers and zero skipping logic[7].

In this evaluation, we use IDW problem matrices emanating from automatic MVS. The results represent averages over ten frames of ten S3D video sequences from [155]. The evaluated matrix sizes are $M = \{8\,\mathrm{k}, 32\,\mathrm{k}, 128\,\mathrm{k}\}$, and correspond to LS problems defined on 64×128, 128×256 and 256×512 grids using a 9-stencil. Precision-related issues are discussed next, followed by a discussion of the main results.

4.6.1　Matrix Condition and Solution Accuracy

The matrix condition considerably influences the convergence and solution accuracy of all solvers [136–138]. In the problem at hand, the condition is mainly influenced by the amount of data constraints in the LS problem. As shown in Section 4.3.5, the more data constraints are available, the lower is the condition number and the better is the solvability. Intuitively, this is because the LS problem propagates the data constraint values from the constraint positions to adjacent

---

7. We base this overhead on our practical experiences gained during the design and implementation of several such systems in both ASICs and FPGAs.

variables via the differential smoothness constraints. The less data constraints are given, the further these values have to be propagated and the more susceptible the solution process is to quantization.

In order to assess the impact of the matrix condition, the dataset used in this evaluation comprises a range of matrices with low to very high condition numbers (see Figure 4.2c). The arithmetic precision (number of mantissa bits) and additional parameters such as the tolerance ϵ of the PCG solver, and the threshold ρ of the ICHOL solver are determined for each matrix size separately using the worst-conditioned problem in the dataset. Note, that the solution $\mathbf{f}$ represents pixel coordinates and therefore it is sufficient for the maximum absolute- and median absolute error to lie below 1 and 0.1 pixel, respectively (smaller errors are imperceptible). The chosen parameters of all evaluated solver variants are summarized in Table 4.1.

4.6.2 Results

The area time energy (ATE) design space of all evaluated variants is shown in Figure 4.8, and Figure 4.9 shows additional information such as the external bandwidth, and a more detailed energy and area split in the case of 128×256 matrices. We can see that the LDL[on] and ICHOL[on] variants are only viable for small 8 k matrices, and otherwise they are suboptimal due to the increasing size of the matrix decomposition $\mathbf{L}$. The PCG[off] variant has low area requirements, but apart from this, it is quite inefficient from an energy and throughput point of view. This is due to the many off-chip memory accesses, which causes the solver to be completely memory bound (as can be seen in Figure 4.9a). The ICHOL[off] variants can reach high energy efficiency comparable to the PCG[on] solvers, but they cannot be scaled as well as PCG due to data dependencies. A PCG solver that uses only on-chip resources can provide good energy efficiency at moderate area occupation in all cases – provided that the matrix is well-conditioned. A similar observation can be made for the ICHOL[off] variants, although they do not reach the same level of energy efficiency as PCG[on].

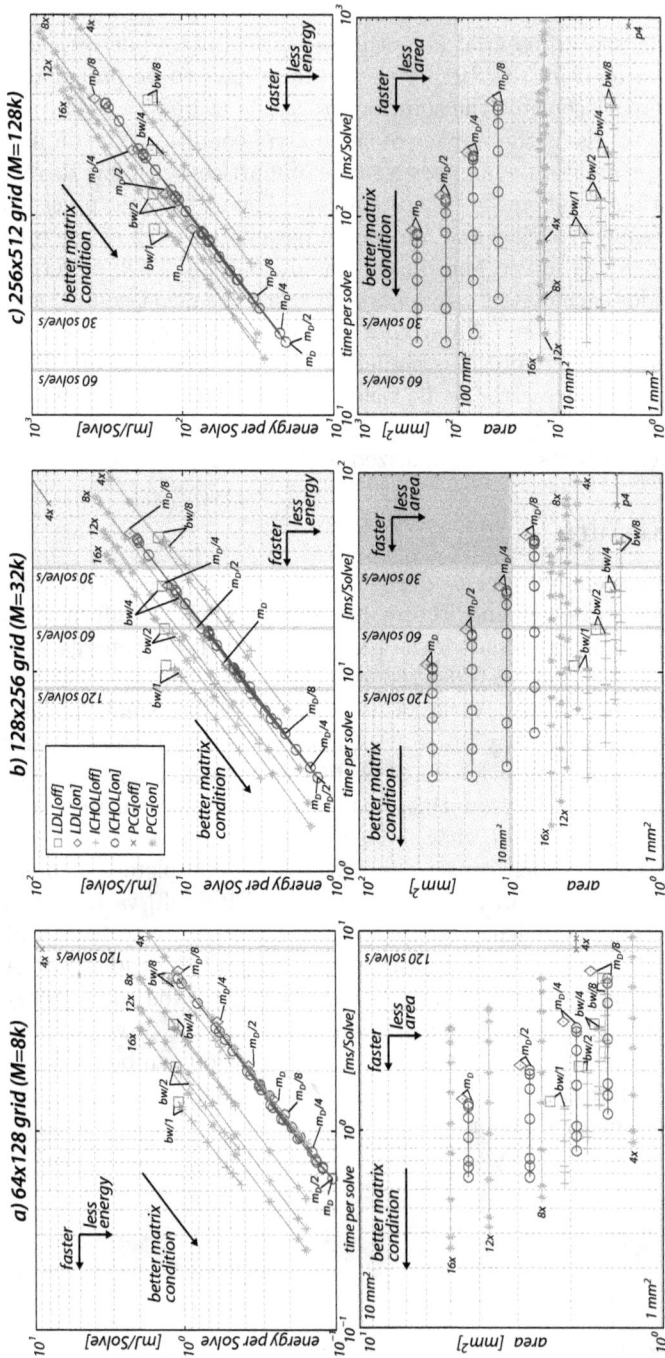

FIGURE 4.8: This figure shows the ATE design space of all evaluated architecture variants for the grid sizes a) 64×128, 128×256 and 256×512. Areas which are deemed infeasible due to an insufficient throughput or a very large area requirement are shaded in red.

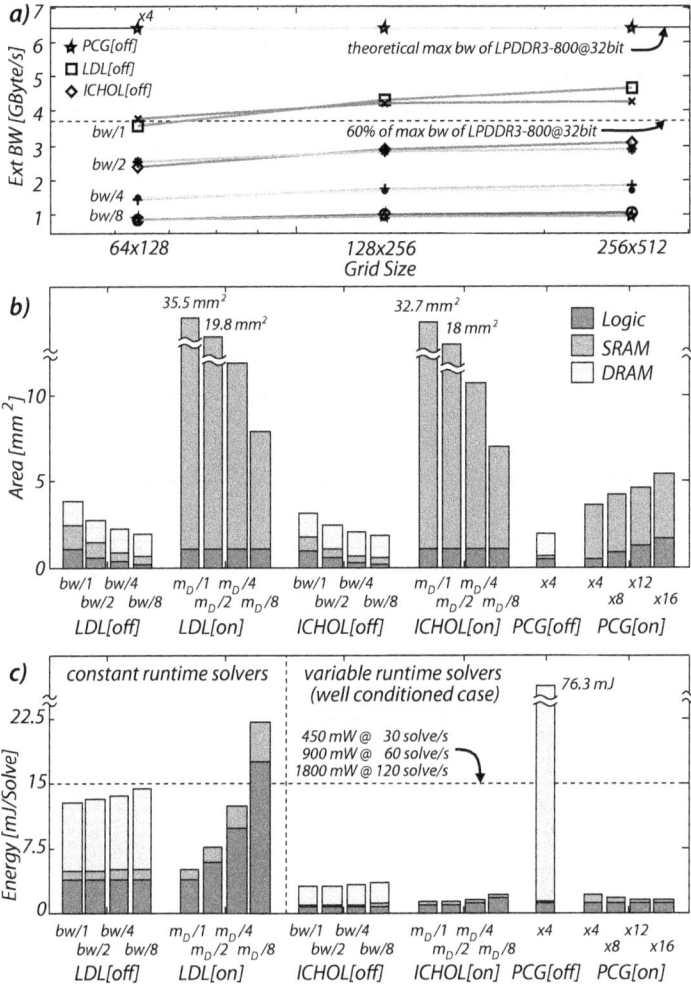

FIGURE 4.9: a) External bandwidth for all off-chip variants for well-conditioned problems. 100% and 60% of the maximum theoretical bandwidth are indicated with a solid black and a dashed black line. The latter is a more realistic estimate for the maximum bandwidth available in a real system. In b) and c) an area and energy split is shown for 128×256 grids. For comparison, the GPU subsystems of APPLE's A7 and NVIDIA's Kepler are estimated to occupy around $21\,mm^2$ and $60\,mm^2$, respectively [22].

Direct solvers such as ICHOL[off] and LDL[off] are suitable for grid sizes is in the order of 128×256 to 256×512 if circuit area is of utmost importance, and if a memory interface with sufficient off-chip bandwidth (Figure 4.9a) is available. On the one hand, the ICHOL[off] variants have the advantage of requiring less energy per solve. On the other hand, the LDL[off] variants are conceptually simpler and have a constant runtime, which is an important aspect for real-time implementations.

4.6.3 Discussion

There are two main usage scenarios for the accelerators at hand: either they are singled out into a separate chip, or they are integrated as a part of an SoC. In both scenarios, a DRAM free solution is highly desirable. In the first case, the chip would have much simpler I/Os and packaging, and it minimizes system cost and board level design effort. In the SoC scenario, a DRAM controller is probably available, but DRAM bandwidth will be in high demand due other subsystems (CPU, GPU, DSP...). So adding an additional DRAM-bandwidth hungry accelerator is a difficult proposition. As can be seen in Figure 4.9a, all off-chip variants use a significant percentage of the available external bandwidth in the evaluated scenario. An accelerator that fully works in a streaming-like manner is much easier to insert at the SoC level – like an ISP that is used to process the camera output in modern SoCs.

From this perspective, the iterative PCG[on] solver variant is highly attractive since it does not require as much silicon area as the LDL[on] variants, and since it is easier parallelizable – which is important to support larger matrix sizes in the future. Issues due to the matrix condition are fortunately less severe than the results might indicate, since the matrices are often well-conditioned in video processing applications such as IDW (Section 4.3.2). Further, the iterative PCG solver can make use of additional improvements such as using the previous solution as an initial guess and more elaborate pre-conditioners [133, 137, 138, 156].

4.7 Summary and Conclusions

The following points are the main takeaways of this chapter:

- We developed a basic estimation framework to rapidly compare and evaluate linear solver architecture variants for IDW problems in terms of energy, circuit area, throughput and precision. The estimation framework is based on a combination of analytic architecture models, pre-characterized logic instances, SRAM datasheet values, and a LPDDR 3 system power model. It could also be used to explore other types of hardware architectures by adding suitable architecture models.

- In area constrained settings, we have seen that direct LDL or ICHOL solvers are the preferred solution in most evaluated cases. In addition, an LDL-based solver exhibits constant runtime, which is a desirable property for real-time applications. Performing an incomplete factorization can reduce the computational complexity, at the cost of more control logic and a variable run time.

- If enough silicon area is available, an iterative solver with on-chip memory such as PCG[on] provides the best tradeoff at the system level in this setting – given that pre-conditioning issues are properly addressed. With a complexity of around $16\,\mathrm{mm}^2$, a PCG[on][16×] solver for 128 k variables is completely feasible at the 28 nm node.

- Direct solvers exhibit many data-dependencies and therefore offer only limited parallelization opportunities. In addition, the memory required for the decomposition data grows super-linearly with the amount of variables. Therefore, iterative solvers are the way to go when problems are scaled up since the operations can be better parallelized, and the storage grows linearly with the amount of variables to be solved.

For larger grids with more than ~128 k variables, however, the evaluated solver architectures do not offer enough throughput for real-time operation, and better algorithms, architectures and implementation technologies are required. To this end, the following points should be considered.

- The bottleneck for iterative algorithms is their high memory bandwidth, which becomes a burden when scaling to larger matrices as either a large on-chip SRAM or a high-bandwidth off-chip DRAM is required. New memory technologies [157] such as 2.5D integration or 3D stacking have the potential to remedy this issue. For example, emerging FPGA devices like the UltraScale+ series from XILINX [158] will contain significantly more on-chip SRAM with of capacities up to 500 Mbit[8].

- With larger matrices, numerical issues may arise. In such cases it is advisable to evaluate CG variants such as CGLS and LSQR [148,149]. Algorithms developed for symmetric indefinite matrices (such as minimum residual (MINRES) [159]) are also worth evaluating, since they may converge faster than CG in some cases [160].

- When problems are scaled, elaborate preconditioners become essential for rapid convergence. One of the key properties of such preconditioners is that they should be well parallelizable. For instance, incomplete factorizations (such as ICHOL) have the potential to significantly accelerate PCG for positive definite problems – yet, the convergence gains cannot always be realized due to the sequential nature of the forward/backward substitution steps which have to be performed for each preconditioner solve. A careful evaluation of existing parallel pre-conditioning techniques [134,156] should therefore be carried out. Also, iteration concepts such as multi-grid solvers [161] and preconditioned variants thereof [133] have recently shown promising results in the context of computational photography.

- Further, we have seen that approximations such as an ICHOL can boost the performance of direct solvers. Approximation concepts for iterative solvers should therefore also be evaluated in

---

8. For comparison, the ALTERA Stratix IV device used to implement the MVS prototype in Chapter 7 only contains around 20 Mbit of on-chip SRAM, although it is one of the biggest FPGAs of that series.

more depth. While there already exist many such techniques on the algorithmic level [136–138, 156], it would also be interesting to look into approximations on the micro-architectural- or on the circuit level [162], since iterative solvers are self-correcting to some degree.

- As mentioned previously in Chapter 3, a more radical approximation method that could be evaluated in more detail are EA diffusion filters. Such filters scale better with the amount of variables, and their mathematical properties allows to calculate fast approximations to regularization problems. An alternative way would be to use them as some sort of preconditioner for iterative solver methods.

Based on the evaluation performed in this chapter, we decided to implement an optimized instance of a direct solver in the MVS prototype in Chapter 7. There are several reasons for this decision. First of all, the target platform is an ALTERA Stratix IV FPGA where on-chip memory resources are limited and have to be shared among several blocks. Further, a DRAM memory was readily available on the employed FPGA platform. In such a setting, and for problems with around 20 k variables, the direct solver variants LDL and ICHOL with off-chip memory are clearly the best choices. In this case, we therefore decided in favour of the LDL[off] solver due to its simplicity and constant runtime. The ICHOL[off] variant would dissipate less energy than LDL[off], at the cost of a variable runtime and a conceptually more complex datapath. It should be noted however, that this choice was made for this specific demonstrator system. For a newer FPGA or a completely integrated SoC solution, a PCG[on] solver is probably the better choice due to the higher energy-efficiency.

Chapter 5

Efficient Resampling for MADs

5.1 Introduction

As an alternative to stereoscopic TV displays which require glasses to perceive the 3D effect, MADs have recently become available on the consumer market. These displays offer a glasses-free 3D experience, and a limited motion parallax in horizontal direction by projecting several views of the scene into different spatial directions. One key challenge of MADs is to provide these multiple views to the viewer. Beside the technical difficulty of acquiring MV content, the number of views is display-specific and can currently range from five to more than twenty views. Thus, transmitting and storing MV is often infeasible and automatic MVS algorithms have emerged as a viable alternative.

Any MVS algorithm involves a non-linear image transformation step. Although each output view can be treated as a separate resampling problem, MV displays require adapted anti-aliasing filters as described by Konrad et al. [5, 163–166]. In this chapter, we provide a comparison of several resampling approaches[1] in terms of visual artifacts and computation costs (#operations and memory requirements).

---

1. Note that we only address aliasing artifacts due to the non-orthogonal sampling pattern in this work. There exist other filtering techniques [5,6] that also address crosstalk issues between the views due to light leakage in the lenticular sheet.

The evaluation primarily targets IDW-based MVS methods Stefanoski et al. [14], but the results can also be applied to DIBR methods [13]. While a large body of previous work is concerned with single image resampling [15,167,168], their findings cannot be directly extended to MV sampling.

5.2 Background

In this work, we analyze the following model for MV display resampling. High definition S3D video is converted to M output views (Figure 1.1), using a non-linear forward warp function that maps each input view to a transformed output view. The warped output views are then resampled to form an *interleaved* MV image. In the following, we summarize these steps in more detail, based on [5,15,163–168].

5.2.1 Ideal Single View Resampling

An image warp is defined as continuous mapping $\mathbf{x} = \mathbf{m}(\mathbf{u})$ that maps any coordinate $\mathbf{u} \in \mathbb{R}^2$ from the input domain to a location $\mathbf{x} \in \mathbb{R}^2$ in the output domain. In order to transform a discrete input to a discrete output, four steps are generally employed. First, a continuous signal $f_{\text{in}}^{\text{C}}(\mathbf{u})$ is reconstructed by applying an interpolation filter h_{i} to the discrete input. Second, the continuous signal is warped to $f_{\text{warp}}^{\text{C}}(\mathbf{x})$ using $\mathbf{m}(\mathbf{u})$. In a third step, an anti-aliasing filter h_{aa} is applied to achieve an aliasing-free output signal $f_{\text{aa}}^{\text{C}}(\mathbf{x})$. Finally, this signal is sampled to form the discrete output image. See Figure 5.1 (red arrow) for an illustration.

5.2.2 Sampling for Multiview Displays

Image sampling is usually performed using orthogonal sampling grids. In contrast, MADs often employ slanted lenticular lenses to achieve view-multiplexing [5], and thus result in non-orthogonal or irregular sampling grids (Figure 5.2). Unfortunately, classical sampling theory cannot be applied to define the anti-aliasing filter h_{aa} in such cases.

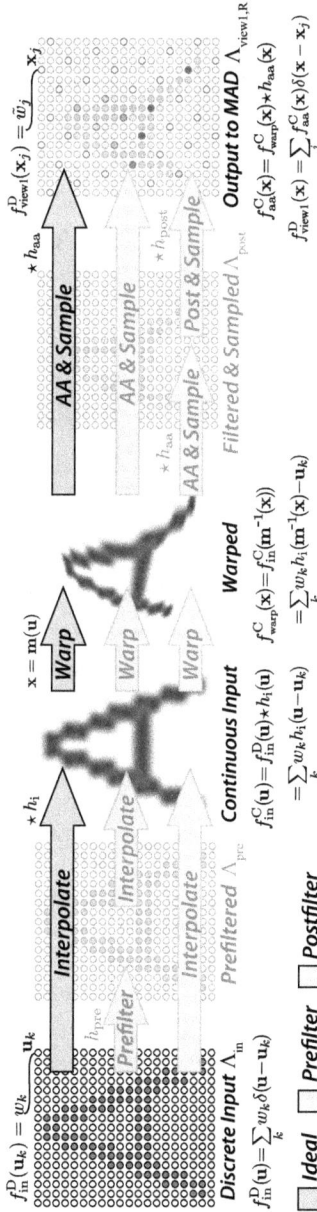

FIGURE 5.1: MV resampling: the red arrows (top) reflect the steps of the analytical resampling problem as presented in Section 5.2. The green arrows (middle) and blue arrows (bottom) indicate the steps of practical implementations using *pre-filtering* and *post-filtering*, as presented in Section 5.3.2. Note that the MV output view is sampled on a non-orthogonal sampling lattice.

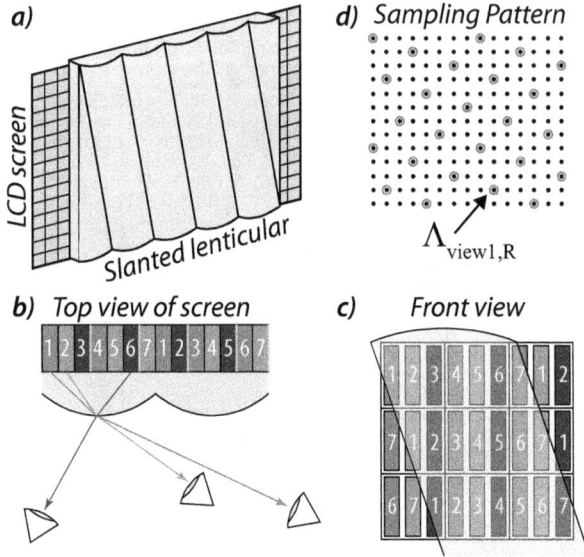

FIGURE 5.2: Lenticular sheets are typically mounted in a slightly slanted arrangement on the MV screen in order to reduce Moiré artifacts and balance horizontal and vertical resolution of the sub views (a). The lenticular sheet deflects the light emitted by different subpixels into different spatial directions (b). The slanted arrangement (c) leads to non-orthogonal sampling lattices (d). The illustrated sampling pattern is for the ALIOSCOPY HD displays [169]. $\Lambda_{view1,R}$ denotes the red channel of the first view.

As shown by Konrad et al. [163–166], an ideal filter needs to limit the frequency content to a *primitive cell* of the *inverse sampling lattice*. In general, this results in non-separable anti-aliasing filters with an intricate shape. For example, the lattice in Figure 5.2 d) leads to a filter with a hexagonally shaped frequency response (shown in orange in Figure 5.3a). However, for natural images, this ideal filter can be approximated by using more efficient separable filters [165].

5.2.3   Resolution Ratio

Due to the view-multiplexing of MADs, the resolution effectively used of each input view is often considerably smaller than the overall resolution. More formally, a *resolution ratio* between the input images and the interleaved output image can be defined as the quotient of the density of their sampling lattices [163]. This ratio is denoted as $\rho(\Lambda_1, \Lambda_2) = d_2/d_1$ in the following, where d_1 and d_2 are the corresponding densities of the lattices Λ_1 and Λ_2. The ratio allows to describe a change in resolution for non-orthogonal sampling lattices (e.g. downsampling for $\rho(\Lambda_1, \Lambda_2) < 1$), and plays an important role in our evaluations, as it allows to convey a resolution change (e.g., downsampling) in concise manner – regardless of whether the lattices are orthogonal or not[2].

5.3 Practical Multiview Resampling

Having introduced the general steps of MV resampling, we now address the problems that arise with computationally efficient filter kernels. We then show different strategies for MV rendering and highlight their potential pitfalls.

5.3.1 Resampling Artifacts

Generally, there is no closed-form analytical expression for the rendering equation $f_{\text{view}}^{\text{D}}(\mathbf{x})$ useable in practice. Thus, it has to be approximated with non-ideal filters and intermediate rasterization steps, which in turn gives rise to blur and aliasing. Blur is due to attenuation in the filter pass band and can be reduced by using higher order filters. Aliasing is due to insufficient attenuation in the stop band which causes components of the periodic spectrum of the input image to alias back into the pass band upon re-sampling. We will focus on aliasing artifacts because aliasing is specifically noticeable in MV resampling applications due to non-orthogonal sampling grids.

There are two potential sources for aliasing: the non-orthogonal display sampling and the warp transformation. To analyze aliasing artifacts due to non-orthogonal sampling, we assume a simplified setup with unit mapping $\mathbf{m}(\mathbf{u}) = \mathbf{u}$ and similar sampling densities $\rho(\Lambda_{in}, \Lambda_{view}) \approx 1$. In this case, both filters h_i and h_{aa} in Figure 5.1 (red arrows) can be combined into one filter $h_{i\_aa} = h_i \star h_{aa}$. When a

---

2. To make the distinction more concise, we use the term 'grid' only for orthogonal sampling arrangements, and the term 'lattice' for arbitrary arrangements.

low-order separable filter such as the bilinear kernel is used, aliasing can appear due to the mismatch to the bandwidth shape. While this effect is usually negligible for orthogonal sampling grids, it is much more visible for non-orthogonal lattices. As the spectral components of the input are also replicated non-orthogonally, these replicas can bleed into very different regions of the hexagonally-shaped display bandwidth, shown in Figure 5.3 a-b. This results in aliasing patterns with *different directionality*, which can be seen in the reconstructed[3] view Figure 5.3 c).

Aliasing due to the warp manifests itself in similar but more localized artifacts. These artifacts can occur in regions that are strongly contracted. In MVS, this type of aliasing is typically less visible, as image compressions are only performed in less salient regions such as backgrounds or uniform regions [14]. Furthermore, in S3D applications, transformations are mostly restricted to horizontal translations since the transformations originate from *horizontal disparities* [2].

5.3.2   Removing Aliasing Artifacts

There are two distinct ways to mitigate aliasing due to non-orthogonal sampling. First, the spectral copies can be attenuated better by using a higher order filter h_{i_aa}, as shown in Figure 5.3 c-e). Second, the spectral copies can be spaced further apart by using oversampled input images, as shown in Figure 5.4 a-f). If a separable, global scaling transformation is employed, both the interpolation h_i and anti-aliasing filter h_{aa} can be integrated into the resampling pipeline easily in one single step. However, when using arbitrary transformation mappings $\mathbf{m}(\mathbf{u})$, combining h_i and h_{aa} becomes computationally challenging, as most filter kernels cannot be transformed analytically.

Thus, one solution is to perform the interpolation with a low order interpolation kernel h_i and apply a high quality *post-filter* h_{post} to the transformed image (the technique is illustrated in Figure 5.1 with blue arrows). This post-filter is then independent of the transformation and can be implemented efficiently. Aliasing artifacts due to the warp

---

3. The display views are reconstructed on a rectangular 1080p lattice with a high order (51×51), hexagonal finite impulse response (FIR) filter in order to simulate the MAD-screen-to-eye transfer function. The filter was designed with the fwind2 method from MATLAB's image processing toolbox (using a Hann window).

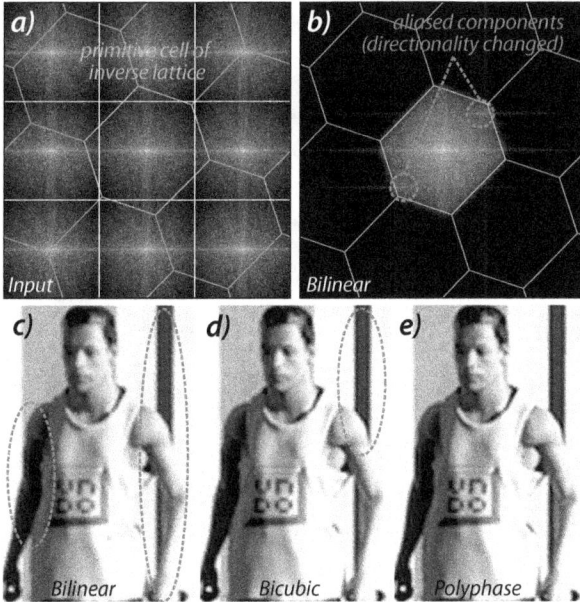

FIGURE 5.3: Aliasing artifacts due to non-orthogonal view sampling: a) shows the periodic input spectrum, where the white boxes illustrate the input bandwidth. The bandwidth of one view of the Alioscopy HD 42" display [169] is illustrated with the hexagonal cells (the primitive cell is drawn in orange). c-d) show reconstructed display views for different resampling filters of increasing complexity: c) bilinear, d) bicubic, e) 6×6 polyphase Lanczos (use your document viewer to zoom in, in order to better see the artifacts). b) shows the resulting spectrum of c). The frequency components in the red circles cause aliasing artifacts with changed directionality.

are barely visible in this setting, as no directionality change happens when the transformed image is sampled on the orthogonal grid Λ_{post} (Figure 5.5a) shows a contracted image region that was rendered with a post-filtering method and no anti-aliasing filter h_{aa}). Anti-aliasing is not strictly required, but it could be performed by adaptively supersampling the output image on a higher resolution along the horizontal direction, for example.

Using the fact that for stereo applications the warp is approximately a global translation, i.e., only has small local variations, we propose a second strategy that moves the high-quality filter to the

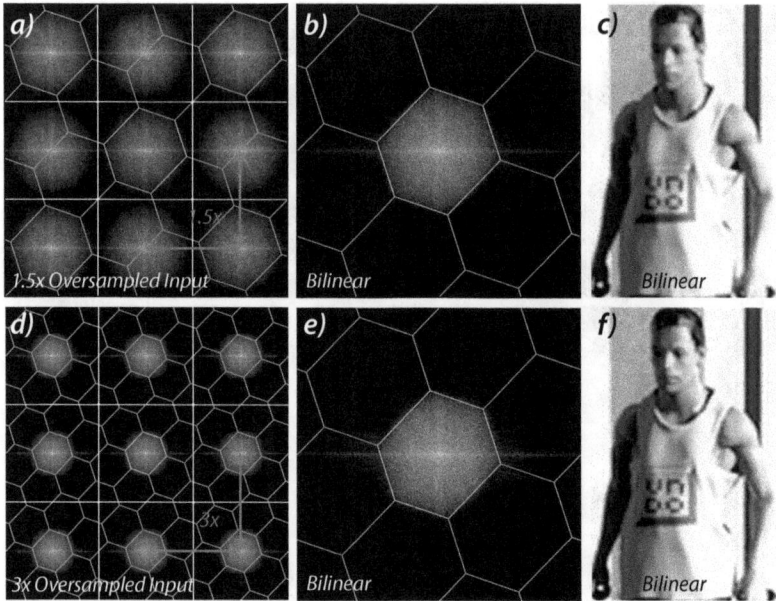

FIGURE 5.4: Oversampling can be used as an alternative to high order filters to reduce aliasing. a-c) show the effect for 1.5× and d-f) for 3× oversampling, using a separable 6x6 polyphase Lanczos kernel to upsample the input image, and a simple bilinear filter to perform the actual resampling.

input image domain. That is, we use oversampling in the input domain, by applying a high quality *pre-filter* h_{pre} to up-sample the input image (illustrated in Figure 5.1 with green arrows). Deviations from a global translation in the warp now causes aliasing artifacts that exhibit a directionality change since the transformed image is sampled on the non-orthogonal grid Λ_{view}. Thus aliasing due to the warp more visible than with post-filtering is in this case (Figure 5.5b). The effect can be mitigated using adaptive supersampling or by using adaptive splatting [50] as shown in Figure 5.5 c-d). The pre-filtering strategy requires significantly less computations and intermediate memory in MV settings where a large amount of synthetic views is generated from a small amount of input views (e.g., $N = 2$ in the case of S3D input).

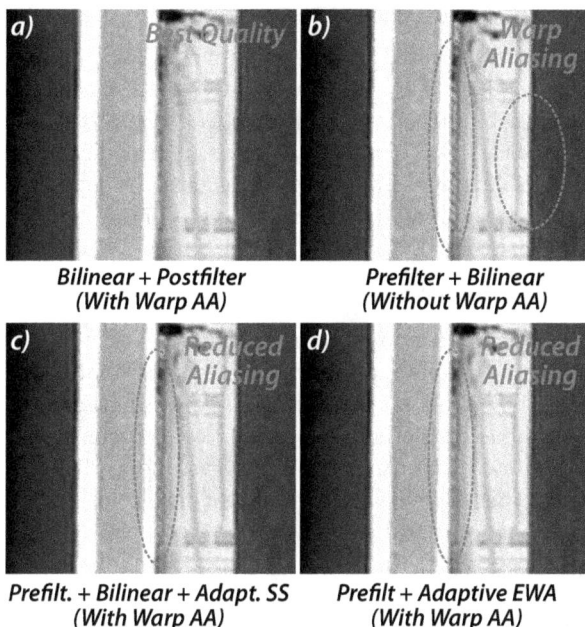

FIGURE 5.5: a) and b) show the effect of post- and pre-filtering with a bilinear kernel h_i and no filter h_{aa}. Aliasing is barely visible in a). c) shows similar results, with additional adaptive $2\times$ supersampling in horizontal direction using a simple box filter h_{aa}. d) shows pre-filtering with adaptive EWA splatting [50], where Gaussian kernels are used for h_i and h_{aa}.

5.4   Evaluation

In this section, we compare the resampling quality and computational complexity of efficient and practical resampling strategies for MV rendering. First, we will describe the evaluation setup. Then, we discuss the evaluated resampling strategies. Finally, we will discuss the benefits and drawbacks of the evaluated algorithms.

PSNR Measurements

In order to assess the quality of different resampling strategies, we measure the mean peak signal-to-noise ratio (PSNR) of one reconstructed view with respect to a high quality reference over 31 frames of the

Undo Dancer test sequence [170]. The output views are reconstructed as described in Section 5.3.1. The reference image is rendered from 1080p input footage with a high quality two-step splatting algorithm and a nonseparable 51×51 post-filter h_{post}. The algorithm employs 6×6 Lanczos kernels for h_i and h_{aa} and an intermediate rasterization step with variable supersampling to transform the input images in a similar manner as proposed by Szeliski et al. [171].

Complexity Estimates

To compare the computational complexity of the different approaches, the number of operations and memory costs are estimated analytically. We assume 10 % of all rendered pixels require anti-aliasing for the adaptive anti-aliasing methods[4]. Memory costs assume that the MV warping functions exhibit a maximum Keystone (y-disparity) of two pixels on the input lattice with the highest resolution Λ_{in1}.

Experimental Setup

Our evaluation is performed using the parameters of the ALIOSCOPY HD displays [169], which support eight views interleaved into a super-lattice with resolution 1920×1080 RGB pixels. As the size of the input views has an impact on computational complexity and image quality, we evaluate each resampling configuration on three different input resolutions, denoted with the three prefixes *r1-r3*:

- r1: 1920×1080, ratio $\rho\left(\Lambda_{in1},\Lambda_{view}\right)=1/8$,
- r2: 960×540, ratio $\rho\left(\Lambda_{in2},\Lambda_{view}\right)=1/2$,
- r3: 640×360, ratio $\rho\left(\Lambda_{in3},\Lambda_{view}\right)=9/8$.

5.4.1 Resampling strategies

In our evaluation, we focus on practical and efficient resampling strategies. More specifically, we compare *bilinear, bicubic and Gaussian*[5] interpolation kernels h_i and their pre- and post-filtering variants. All pre-filtering variants up-sample the input images onto a 1920×1080

---

4. This value is an estimate for typical MVS warping functions and has been obtained by determining the amount of pixels residing in compressed image regions.
5. Gaussians are rasterized inside a bounding box with $\lceil5\cdot\sigma\rceil$ pixels edge length.

grid Λ_{pre}. All post-filtering variants retain the same resolution $\Lambda_{in} = \Lambda_{post}$ before subsampling to Λ_{view}.

The bilinear and bicubic kernels h_i are implemented using backward mapping, and will thus require the inversion of $\mathbf{m}(\mathbf{u})$. The 4 variants will be denoted as *BilPre*, *BilPost*, *BicPre* and *BicPost*. Aliasing due the warping is mitigated using adaptive $2\times$ supersampling in horizontal direction using a simple box filter h_{aa}.

Gaussian kernels allow for efficient forward mapping and do not require the costly warp inversion of backward mappings. Furthermore, Gaussians are closed under affine transformations which allows to analytically combine interpolation h_i and anti-aliasing h_{aa} into one single filter (EWA splatting [36, 50]). The respective pre- and post-filtering methods are denoted as *EwaPre* and *EwaPost*. As additional simplification, the post-filter h_{post} can be chosen as Gaussian and merged with h_i and h_{aa} into one single filter kernel, such that only one filtering step has to be performed. This combined EWA variant will be denoted as *EwaComb*. Note, that neither of the three variants requires super-sampling, as forward EWA rendering is implemented as an area sampler and thus implicitly includes anti-aliasing.

Warp Interpolation and Inversion

The warp $\mathbf{m}$ is usually not provided as analytical function but described by samples of a discretized grid. In the application at hand, [14] this mesh contains forward mapping coordinates. Furthermore, it is often not provided at pixel-resolution but approximated at lower resolution, usually at $10\times$ smaller widths and heights. This has implications for the resampling strategies employed, which operate on pixel-dense coordinate information. In this evaluation, we assume that forward mapping methods bilinearly upsample the warp coordinates. Backward-mapping methods have to additionally invert the quadrilateral warp mesh for every output sample. Note, that higher-resolution meshes require less intermediate values to be stored in memory, but also result in higher computational complexity. In our evaluation, this warp inversion is performed using Barycentric coordinates [76] for performance reasons, however, this may lead to visible artifacts for coarser mesh sizes. The complexity of backward mapping methods is thus reduced in applications using the inverse map $\mathbf{m}^{-1}(\mathbf{x})$.

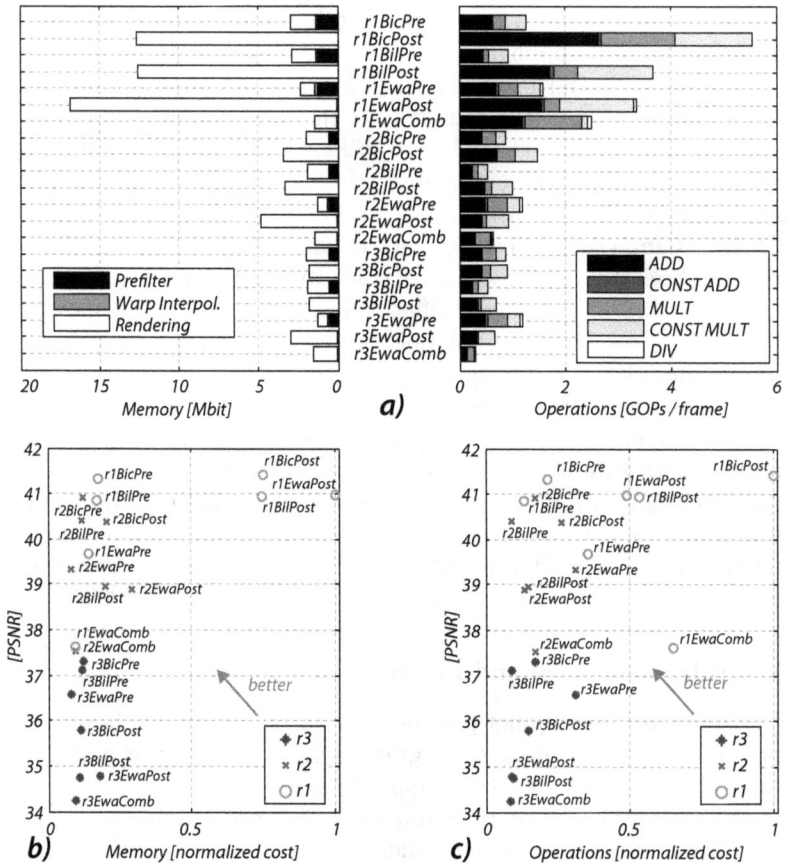

FIGURE 5.6: a) shows estimates of the computational cost of the evaluated methods. b) and c) show the operation and memory costs (normalized) versus image quality (mean PSNR value over 31 frames of our test sequence, evaluated on one view). The PSNR is calculated w.r.t. a high quality reference image (Section 5.4). The operation cost is calculated by relative weighting of the operation types. The weights were estimated by synthesizing 20 bit fixed-point operators in 65 nm CMOS technology at 400 MHz: add = 0.057, add const = 0.043, mult = 0.286, mult const = 0.093, div = 1.0.

5.4.2   Discussion

Figure 5.6 a) shows the estimated complexity of the evaluated algorithm variants and b,c) relate those estimates to the image quality. As can be seen from the PSNR plots, rendering from higher resolution (or oversampled) input footage leads to significantly better quality, in general. However, all post-filtering variants suffer from increased operation and memory cost in this case. This becomes even more pronounced for a bigger number of views, as all intermediate results Λ_{post} need to be stored and downsampled. In contrast, all pre-filtering variants offer considerably lower complexity when rendering from higher input resolution, as only a small amount of output pixels have to be super sampled.

The advantage of all EWA variants is that they do not require warp-inversion, which makes them conceptually simpler than backward mapping methods in this setting. Also, since EWA is an area sampler, these variants can inherently handle warp anti-aliasing better. However, they show a tendency to produce low PSNR values since Gaussian filters are not strictly interpolating like bilinear or bicubic interpolation. Therefore PSNR saturates around 37.6 dB even for high resolution footage for all EWA variants.

The combined EWA filtering approach *EwaComb* offers an interesting design alternative for r1 and r2 footage. Complexity-wise, it is between pre- and post-filtering methods, and it offers relatively low memory cost. The somewhat lower PSNR values are mitigated by the fact that it does not require warp-inversion and offers the best-warp anti-aliasing capabilities. However, it should be noted that it does not perform well with low-resolution input (r3), as the low filter order leads to visible global stripe artifacts due to the display sampling.

5.5 Summary and Conclusions

Our evaluations show large differences in computational complexity, memory requirements and visual quality among the evaluated algorithm configurations. In particular, we can draw the following conclusions:

- The highest visual quality is achieved by *post-filtering* variants using high resolution input images. However, this approach is computationally expensive, requires a lot of memory, and therefore does not scale well with the number of views.

- All EWA splatting variants result in slightly lower PSNR values than the bicubic and bilinear filters due to the properties of Gaussian filters, but have the conceptual advantage of not requiring warp inversion. Also, these methods offer the best warp anti-aliasing capability. Instead of applying a pre- or post-filter, the display anti-aliasing filter and warp anti-aliasing filter can be merged into one combined Gaussian kernel (*EwaComb*). This represents a conceptually even simpler design alternative. However, it does not result in the best complexity and quality tradeoff.

- The *pre-filtering* variants with bicubic or bilinear filters and 1.5×-3× oversampled input images offer the best trade off between image quality and computational complexity – regardless of the input resolution. They also show good scalability with respect to the number of views, since only the required display pixels are calculated and the high quality filtering is performed on the input images.

Other forward mapping algorithms optimized for 1D shift transformations, such as the method by Berretty et al. [172] for example, might be a better alternative than EWA splatting for the specific application of MV rendering. However, EWA is more general and therefore compatible with other IDW applications requiring 2D transformation, such as, e.g., video retargeting.

The 'Madmax' ASIC [36] used in the rendering stage of the MVS prototype (Chapter 7), was developed prior to this work, and implements the combined EWA splatting variant termed *r1EwaComb* in this evaluation. This resampling strategy has been chosen due to its conceptual simplicity (no warp inversion, no additional filter stages), and since it offers good warp anti-aliasing properties. It is therefore able to also handle other IDW applications with more pronounced deformations, such as video retargeting. However, the evaluation conducted suggests that a more efficient (and more specialized) architecture can be designed by switching to one of the pre-filtering strategies proposed in this chapter.

Chapter 6

Logarithmic Number Units

6.1 Introduction

Energy-efficient computing and low-power operation are strong requirements for a vast number of application areas ranging from high-performance computing (HPC) over mobile consumer electronics to embedded systems and devices for the internet of things (IoT). While for some applications integer and fixed-point arithmetic suffice, several others[1] require a larger dynamic range, typically obtained using single-precision FP numbers. In addition, many algorithms are initially developed for general-purpose computing systems (PCs, workstations) assuming HDR arithmetic. Porting these algorithms to integer or fixed-point arithmetic is a labor-intensive and technically challenging task which requires extensive testing and verification [177]. Hence, there is a trend toward supporting HDR arithmetic, currently in the form of single-precision FP, in computing fabrics such as FP-GAs [178] and even on low-power microcontrollers such as the ARM Cortex M4 [179].

---

1. Examples range from classification [173, 174], vision applications like detection [175], SLAM [132], camera pose estimation [176], to STEA filters (Chapter 3) and solvers (Chapter 4).

Unfortunately, FP is energy-hungry, and significant research effort is devoted to reducing the energy required for HDR computing. To this end, the LNS has been proposed [180–189] as an energy-efficient alternative to conventional FP, since several complex operations such as MUL, DIV, and SQRT translate into simple integer operations in LNS. This is not only relevant for HPC, but also increasingly needed for low-power, low-cost embedded applications where the demand on intensive DSP capabilities continues to grow.

However, the drawback of LNS is that additions and subtractions become non-linear operations and, when implemented in hardware, have to be approximated with a dedicated LNU. The area of LNUs grows exponentially with the desired precision. For an accuracy equivalent to single-precision FP, LNUs are typically larger than traditional floating point units (FPUs), which makes it difficult to justify their use in general purpose processors.

When viewed in a multi-core setting, the area overhead of the LNUs change. In such a setting, one LNU can be effectively shared between several processing cores, as the percentage of ADD/SUB instructions in even the most intensive computations usually remains below 30%. This arrangement is more efficient than sharing standard FPUs, as the MUL/DIV instructions in the logarithmic domain can be performed within the integer cores, allowing M DIV/MUL operations to be performed in parallel when only one LNU is shared among M processor cores. Another remedy is to relax precision requirements since certain applications, such as image or audio processing, are error tolerant to some degree and usually do not require single-precision accuracy. Using approximate computing techniques on the architecture and circuit levels [162, 190, 191], significant area and energy savings have been reported with only modest impact on quality.

In this work, we combine these ideas of LNU sharing and approximate computing in order to reduce the LNU area and latency while improving its utilization and overall energy efficiency. We develop a methodology to generate accurate and approximate LNUs capable of natively evaluating LNS ADD/SUB and typecasts (Integer to float – I2F, Float to Integer – F2I), and provide a design space exploration of LNUs in the accuracy range between half- and single-precision FP. Relaxing precision of just a few units in the last place (ULPs) already leads to significant reductions of the LNU area, with insignificant degradation of the output quality when applied to several image and audio processing kernels. Leveraging the similarity of the LNU with special function units (SFUs), typically present in today's GPUs [192], we enhance our architecture to implement transcendental functions with minor area overhead. Our extended LNU supports fused multiply/divide-add operations (FMA, FDA) and the non-linear function intrinsics 2^x, $\log_2(x)$, $\sin(x)$, $\cos(x)$ and atan2(y, x). As an example, these are useful for geometry and phase computations present in many embedded and mobile applications ranging from computer vision [176] to power electronics [193]. A single-precision LNU generated with our framework has similar area compared with the best designs in the literature [187–189], while at the same time providing significantly more functionality. We integrate shared LNUs with different parameters into a multi-core RISC processor cluster in UMC 65 nm low leakage (LL) CMOS technology and present comprehensive simulation and measurement results of benchmark applications from different signal processing domains. Using these designs, we show that for typical non-linear processing tasks, our LNS design can be up to 4.1× more energy-efficient than a private-FPU design and achieve similar energy efficiency when running pure linear algebra kernels with many ADD/SUB operations.

Sections 6.2 and 6.3 give a short introduction to LNS and related work. The architecture and generator framework is described in Sections 6.4 and 6.5. Core integration aspects are covered in Section 6.6, the design space exploration is presented in Section 6.7. The comparisons with related work and the measurement results are given in Section 6.8, while Section 6.9 concludes this chapter.

6.2 Related Work

The LNS has been proposed as a replacement for standard fixed-point and FP arithmetic already in the 1970's [180, 181]. The main challenge of finding efficient approximation methods to implement the non-linear ADD/SUB operations has driven research in the LNS domain. In early papers, implementation of LNUs with accuracy higher than 12 bit was considered infeasible due to exponentially increasing LUT sizes. Several improved implementations have been proposed since. In the low-precision FP calculation domain, with word widths lower than 16 bit, so-called multi-partite table [194] and high-order table based methods [195] have been shown to be effective [196]. LNS based operations have been used to replace fixed-point operations in several applications such as QR decomposition [197], non-linear support vector machine (SVM) kernel evaluation [173], embedded model predictive control [198], neural network accelerators [199] and low power digital filtering [200]. LNS numbers have also been extended to be used for complex numbers [201] and quaternions [202]. Attempts to combine both the advantages of standard FP and LNS into hybrid systems have been made in [203], where the main drawback is the cost of non-linear typecasts.

Coleman et al. [184] introduced the concept of a *cotransformation* to alleviate approximation difficulties related to SUB operations where the result is close to 0.0 (and hence close to $-\infty$ when represented in LNS). As explained in more detail in Section 6.3.4, such cotransformations are analytical decompositions of the problematic functions allowing to implement the same functionality with significantly smaller LUTs. Following the example of Coleman et al., several different cotransformation variations have been presented in [53, 182, 183, 185–189]. Complete LNUs for ASIC processors with accuracy equivalent to single-precision FP have only been presented in [187–189] so far. Coleman et al. [187] describe the *European Logarithmic Microprocessor* (ELM), the first single-core microprocessor featuring an LNU. Their design combines a custom interpolation scheme with the cotransformation developed by [184]. In [188, 189] Coleman et al. improve their ELM design and propose LNUs with lookup tables small enough to be implemented without read-only memorys (ROMs). These LNU designs are confined to basic LNS ADD/SUB instructions and do not have support for casts. Generators for LNS operators on FPGAs have been proposed in [196, 204–206]. Very competitive operators can be generated with the framework presented by Fu et al. [205] which is based on the cotransformation developed by [182] and minimax polynomials [207].

FPUs for standard FP are designed in [208–211] and often only contain support for ADD, SUB, MUL, casts and FMA operations[2]. Support for divisions is then added in form of SW emulations or iterative HW dividers since single-cycle HW divisions are expensive [209]. On top of the basic algebraic operations, there is a growing need to support HDR computations of common non-linear functions. A significant body of work [192, 212–216] studies the efficient implementation of SFUs for GPUs which implement non-linear functions such as cosine, sine, arctangent, square-roots, etc. in FP. Compared to complete numerical function libraries (e.g., as part of math.h in C++) these special instructions are much faster (a few cycles instead of hundreds), but they do not provide the same level of accuracy. Also, the intrinsics usually evaluate the special functions on reduced ranges (e.g., $[0, \pi/2)$), and have to be wrapped with range reduction routines [217].

In contrast to the work listed above, we combine both the LNU and the SFU into one unit, since they share many architectural properties. A similar framework as described by Fu et al. [204, 205] is established in order to generate exact (≤ 1 unit in the last place (ULP) accuracy) and approximate (≥ 1 ULP accuracy) LNUs for different word widths ranging from half to single-precision FP. Based on the parallel ultra-low power (PULP) system previously developed in [218], we design the first multi-processor with a shared LNU, and show that this can be an energy-efficient alternative to a standard FP design for various non-linear function kernels. PULP is an OpenRISC multi-processor platform without FP support, and we used it as an architectural processor template for our evaluations.

---

2. Below is a summary of the instruction labels used in this chapter:

ADD/SUB	addition and subtraction
MUL/DIV	multiplication and division
I2F/F2I	casts: integer to LNS cast and vice versa
EXP/LOG	base 2 exponential and logarithm
SCA/SIN/COS	preprocessing and interpolation of sine and cosine
ATA/ATL/ATN	pre- and postprocessing, interpolation of the arctangent
TRIG	abbreviation for all trigonometric instructions above
FMA/FMS	fused multiply-add and fused multiply-sub
FDA/FDS	fused divide-add and fused divide-sub
MEX/DEX	fused multiply-exponential and fused divide-exponential
FMX	abbreviation for all fused instructions above

6.3   Preliminaries

6.3.1   LNS Number Representation and Format

Standard FP number systems represent a real number a as

$$a = (-1)^s \cdot m \cdot 2^e \qquad (6.1)$$

where s is the sign, m the mantissa and e the exponent. In LNS, real numbers are represented similarly, but without using a mantissa. That is, the number is only represented by an exponent e which now has a fractional part:

$$a = (-1)^s \cdot 2^e. \qquad (6.2)$$

The encoding used in this work is parameterized with the number of integer w_{int} and fraction w_{frac} bits in the exponent. In this case,

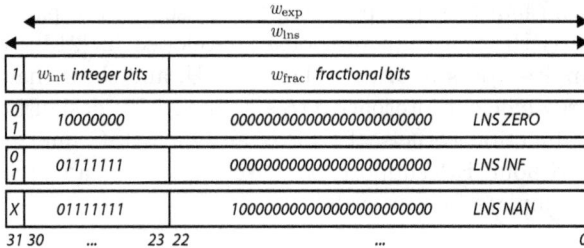

FIGURE 6.1: Encoding of the LNS numbers used in this work.

the exponent is an unbiased two's complement number and its width is denoted as $w_{\text{exp}} = w_{\text{int}} + w_{\text{frac}}$. The word width of the complete number including the sign bit is denoted as $w_{\text{lns}} = w_{\text{exp}} + 1$. For $w_{\text{int}} = 8$ and $w_{\text{frac}} = 23$, the encoding is aligned with the IEEE 754 32 bit single-precision format. Similar to the IEEE 754 standard, special values such as zeros (ZERO), infinities (INF) and not a number (NAN) are encoded using special bit patterns. ZEROs are represented by setting the exponent to the smallest 2's complement value. INFs are represented by setting the integer part of the exponent to the maximum value and the rest of the mantissa to 0. NANs are encoded similarly, but with the highest fraction bit in the exponent set to 1 as illustrated in Figure 6.1 for $w_{\text{int}} = 8$ and $w_{\text{frac}} = 23$.

6.3.2 Arithmetic Operations in LNS

Certain operations can be implemented very efficiently in the LNS. For example, multiplications, divisions, and square-roots can be calculated using a single addition, subtraction or bitshift, respectively.

$$a \cdot b = (-1)^{s_a + s_b} \cdot 2^{e_a + e_b} \tag{6.3}$$

$$a/b = (-1)^{s_a + s_b} \cdot 2^{e_a - e_b} \tag{6.4}$$

$$\sqrt{|a|} = (2^{e_a})^{0.5} = 2^{0.5 \cdot e_a} \tag{6.5}$$

This is an important advantage because calculations are obtained by slightly modified integer arithmetic logic units (ALUs) and result in much shorter latencies than the equivalent FP implementations. However, these simplifications come at the cost of more complex additions and subtractions as these become nonlinear operations in LNS and have to be calculated accordingly:

$$a \pm b = c, \tag{6.6}$$

$$e_c = \max(e_a, e_b) + \log_2(1 \pm 2^{-|e_a - e_b|}). \tag{6.7}$$

Using the absolute difference $r = |e_a - e_b|$, the two nonlinear functions for addition and subtraction can be defined as $F^+(r) = \log_2(1 + 2^{-r})$ and $F^-(r) = -\log_2(1 - 2^{-r})$. These functions are shown in Figure 6.2.

6.3.3 Rounding Modes and Precision

The IEEE 754 standard defines several rounding modes that can be applied after basic arithmetic operations like multiplications and additions. The default rounding mode is *round to nearest*, and provides average and maximum relative errors of 0.1733 and 0.5 ULP, respectively. However, due to the different spacing of the machine numbers in LNS, an ULP in FP is not equivalent to an ULP in LNS. Therefore in [185], Coleman introduced the relations

$$\frac{|\epsilon|_{\text{avg rel arith}}}{2^{w_{\text{frac}}}} = \left(2^{|\epsilon|_{\text{avg log}}} - 1\right) \quad = \left(2^{\frac{|\epsilon|_{\text{avg rel log}}}{2^{w_{\text{frac}}}}} - 1\right) \tag{6.8}$$

$$\frac{|\epsilon|_{\text{max rel arith}}}{2^{w_{\text{frac}}}} = \left(2^{|\epsilon|_{\text{max log}}} - 1\right) \quad = \left(2^{\frac{|\epsilon|_{\text{max rel log}}}{2^{w_{\text{frac}}}}} - 1\right) \tag{6.9}$$

where $|\epsilon|_{\text{avg log}}$ and $|\epsilon|_{\text{max log}}$ are the average and maximum absolute errors in the LNS domain, $|\epsilon|_{\text{avg rel log}}$ and $|\epsilon|_{\text{max rel log}}$ are the average and maximum relative errors w.r.t. to one ULP in the LNS domain, and $|\epsilon|_{\text{avg rel arith}}$ and $|\epsilon|_{\text{max rel arith}}$ are the corresponding relative errors in the FP domain. Using these relations, we can calculate that, e.g. an LNS design with $w_{\text{int}} = 8$ and $w_{\text{frac}} = 23$ should have $|\epsilon|_{\text{max rel log}} < 0.7213$ in the LNS domain in order to have equivalent precision as FP with round to nearest rounding mode ($|\epsilon|_{\text{max rel arith}} < 0.5$).

However, FP equivalent accuracy of 0.5 ULP for a certain word width usually comes at a high cost and is not always required. Hence, it is common to use so called *faithful* designs [196, 205] which deliver a maximum error ≤ 1 ULP. Depending on the definition, an operator is either faithful in the FP or LNS domain (we use the latter in this work). For the distinction between exact and approximate designs, we will use the following definitions. A design for a certain word width configuration $w_{\text{int}}.w_{\text{frac}}$ is considered to be *exact* if its maximum relative error $|\epsilon|_{\text{max rel arith}} \leq 0.5$. A design is considered to be faithful if $|\epsilon|_{\text{max rel log}} \leq 1$ in the LNS domain, and *approximate* otherwise.

6.3.4 Cotransformation

While for low precision implementations with up to around 12 fractional bits $F^{\pm}(r)$ can be stored in LUTs, this approach is not practical for designs requiring higher precision as LUT size grows exponentially with word width. To achieve higher precision, piecewise polynomial approximations have been found to work well [204, 205] – except for operations where r is small since $F^{-}(r)$ has a singularity at zero. This region is termed the critical region and typically ranges from $r \in [0, 0.25)$ to $r \in [0, 4)$, depending on the interpolation scheme employed. In this critical region, so called *cotransformations* [182–184, 186–188] are usually applied in order to decompose $F^{-}(r)$ into

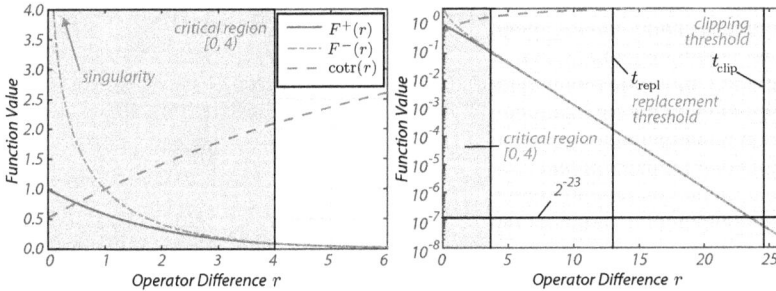

FIGURE 6.2: Plot of the $F^+(r)$, $F^-(r)$ and the cotr(r) functions - note the singularity for $r \to 0$.

sub-functions that can be approximated more efficiently. The cotransformation employed in this work was originally proposed by [182] and decomposes $F^-(r)$ into

$$F^-(r) = -\log_2(1 - 2^{-r}) = -\log_2\left(\frac{1 - 2^{-r}}{r}\right) + \log_2(r)$$

$$= \text{cotr}(r) + \log_2(r).$$

(6.10)

It has been successfully used to create compact LNS operators for FPGAs [204, 205] – but so far its usage for implementing an ASIC LNU design has not been closely investigated. This decomposition leverages the fact that cotr(r) behaves much better around 0.0 as shown in Figure 6.2, and can thus be readily approximated using polynomials. The size of the critical region is set to [0, 4) which provides the best trade-off for this decomposition in terms of overall number of polynomial segments [204, 205]. The $\log_2(r)$ function still has a singularity at 0, but for finite precision arithmetic this function can be efficiently implemented with range reduction techniques of the argument [217], where the argument range can be reduced to [1, 2) by employing a leading zero counter and a barrel shifter. The $\log_2$ function itself can be implemented on this reduced range using a polynomial. Another interesting advantage of (6.10) is the presence of the $\log_2$ term, which can be reused to natively support casts from integers to LNS numbers (I2F) as well as LNS logarithm operations (LOG) directly in HW.

6.4   Architecture Template and Extensions

In this section, we will present the general architecture template of
our LNU. Details on how specific parameters and LUT coefficients are
obtained will be given in Section 6.5. The main design goal of the LNU
architecture shown in Figure 6.3 is to reduce the hardware overhead
and latency. For the targeted accuracy range between half precision
(16 bit) and single-precision (32 bit), 1st and 2nd order minimax polynomial approximations of the functions $F^+(r), F^-(r), \cot r(r)$, and
$\log_2(r)$ have been found to be very efficient and are used throughout
this evaluation. The architecture shown in Figure 6.3 consists of 5
main blocks: the *MulDiv Block*, the *AddSub Preprocessing Block*, the
Main Interpolator Block, the *LogExp Block* and the *Postprocessing
Block*. These are explained in more detail below.

6.4.1   MulDiv and AddSub Preprocessing Blocks

The *MulDiv Block* preprocesses fused multiply/divide-add (**FMA/FDA**)
functions and also enables arbitrary base exponentials $C = \exp(A \cdot B)$
and $C = \exp(A/B)$, termed **MEX** and **DEX**. In addition, the division
capability allows for convenient range reduction of the trigonometric
functions (e.g., division by π for sine and cosine). The intermediate
results U and V from the *MulDiv Block* are used in the *AddSub
Preprocessing Block* which calculates the absolute operator difference
$r = |e_u - e_v|$ and the operator maximum for binary operations such
as **ADD/SUB**. At this point, the block is able to determine which main
datapath units will be activated and generates all control signals for
the LNU, and performs operation dependent preparation steps on the
two operands $U = [s_u, e_u]$ and $V = [s_v, e_v]$. For unary operations such
as **EXP/LOG**, operator V is gated to zero and U is passed through.

6.4.2   Main Interpolator Block

The *Main Interpolator Block* implements $F^+(r)$ on the complete range
$[0, t_{clip})$ and $F^-(r)$ outside the critical region $[4, t_{clip})$ using piecewise
polynomials. Depending on the latency-area trade-off the 1st or 2nd
order approximation will be used. This block is also used for **SUB**
operations in the critical region $[0, 4)$ to evaluate $\cot r(r)$, the result of

which is later added to the $\log_2(r)$ value in the *Postprocessing Block*. For a given input r, the coefficients $p_i^r = p_i(r)$ for $i = \{0, ..., N\}$ (where N is the polynomial order) are selected from a set of LUTs, and the polynomial is evaluated using the Horner scheme

$$p(r) = p_0^r + \delta_p^r \cdot \left(\ldots \left(p_{(N-1)}^r + \delta_p^r \cdot (p_N^r)\right)\right) \tag{6.11}$$

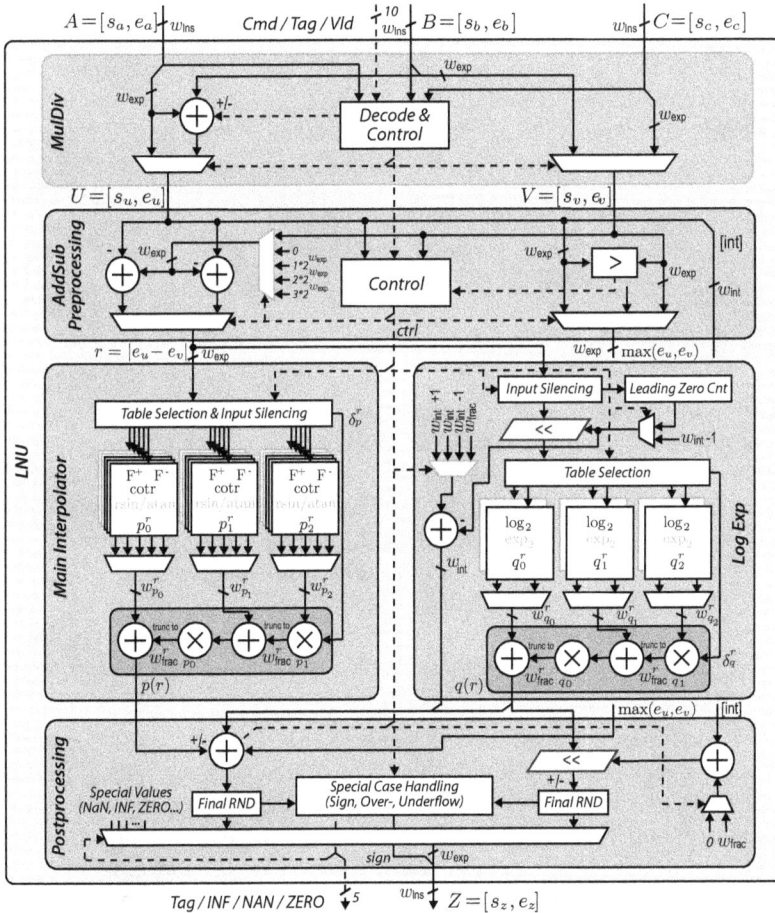

FIGURE 6.3: LNU architecture template used in our generator (shown for interpolation order $N = 2$). Parts for special function extensions (SFEs) are highlighted in orange.

where δ_p^r are the least significant bits (LSBs) of r. The main interpolator datapath can be shared among $F^+(r)$, $F^-(r)$ and $\mathrm{cotr}(r)$. As will be explained in more detail in Section 6.5, each LUT is subdivided into different segments, each of which contains a set of equidistant coefficient samples. The segment boundaries have been aligned to powers of two, such that the segment index can be easily determined by looking at the most significant bits (MSBs) of the argument r. The functions $F^+(r)$ and $F^-(r)$ become increasingly similar as r grows such that one function can be replaced by the other without impact on precision. Therefore, we define a threshold t_{repl} and reuse the $F^+(r)$ tables for $F^-(r)$ when $r > t_{repl}$ (single-precision: $t_{repl} = 14$). Further, for large r, the functions values of $F^+(r)$ and $F^-(r)$ fall below the required precision due to their asymptotic behaviour and can be clipped to 0. This clipping threshold is denoted as t_{clip} (single-precision: $t_{clip} = 24.588$).

6.4.3 LogExp Block

The main objective of the *LogExp Block* is to implement the $\log_2(r)$ function within the critical region $[0, 4)$ required for SUB operations. This is achieved using a barrel shifter and leading zero counter to reduce the range of the input, and an Nth order interpolator with LUTs covering the argument range $[1, 2)$. Note that it is possible to reuse this function for native typecasts from integer to LNS (I2F), and LOG operations in the LNS domain. For a given input r, the polynomial coefficients $q_i^r = q_i(r)$ for $i = \{0, ..., N\}$ are selected from a set of LUTs, and the approximation result $q(r)$ is again calculated as in (6.11).

To natively support inverse typecasts (F2I) and 2^x (EXP) operations, an additional table for the 2^x function has been added. Since this function can also be efficiently implemented using range reduction and polynomial interpolation, we can reuse the existing interpolator to calculate the function value on the range $[0, 1)$, and only have to include an additional shifter at the output which has been moved to the *Postprocessing Block* such as not to increase the delay for the ADD/SUB operations.

6.4.4   Postprocessing Block

The *Postprocessing Block* combines and/or selects the results of the two interpolation blocks. For example, SUB operations in the critical region require the output $p(r)$ of the *Main Interpolation Block* and the output $q(r)$ of the *LogExp Block* to be combined. A final rounding step to the output precision and special case handling such as NAN, over- and underflow detection are also performed.

6.4.5   Trigonometric Functions

Trigonometric functions can be approximated in several ways [215, 216,219], e.g., with the well-known class of CORDIC algorithms [220]. These iterative shift-and-add methods can even be implemented on processors with limited DSP functionality. Lower-latency implementations, however, use table-based methods in combination with range reduction techniques [217]. In fact, several SFUs employ 2nd order interpolation [212–214]. Hence, our LNU can be extended with trigonometric functions by merely modifying a few existing components. We restrict the set of extensions to sine, cosine and arctangent functions, since these are the most commonly used and many other functions can be efficiently derived from these using trigonometric relations (see Section 6.4.5). These functions are implemented using normalized angles which have the advantage of simpler modulo-2 calculations for range reductions (see [215, 216] for more details). We implement the extensions for the functions $\sin(\pi x/2)$, $\cos(\pi x/2)$, $2/\pi \cdot \mathrm{atan2}(y,x)^3$ which already implicitly contain the factor $\pi/2$.

While LNS has several benefits such as low-latency DIV/MUL, it also introduces two difficulties for the particular case of trigonometric functions. First, 0.0 equals $-\infty$ in LNS and leads to singularities as illustrated in Figure 6.4a and Figure 6.5a, where the magnitude of the $\sin(\pi x/2)$ and $2/\pi \cdot \mathrm{atan}(x)$ functions are plotted in the LNS domain. Second, the logarithmic spacing complicates fast range reduction and folding techniques for periodic functions since the quadrants are not evenly spaced. These issues are addressed below for the trigonometric

---

3. atan2(y,x) is a two-argument version of atan(x) available in standard numeric function libraries such as math.h. This version also considers the magnitude and size of the input and returns the angle in the appropriate quadrant.

a) sin(x), double-logarithmic domain

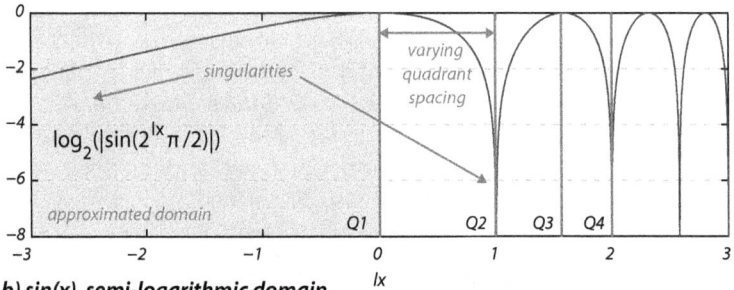

b) sin(x), semi-logarithmic domain

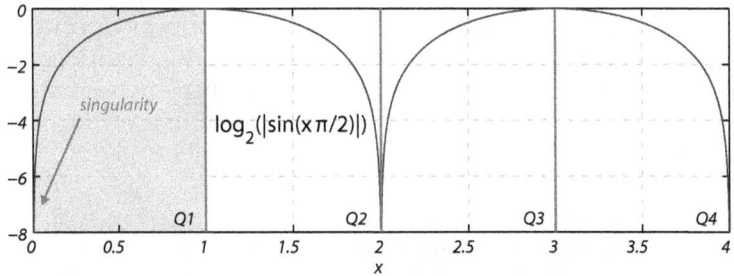

c) rsin(x), semi-logarithmic

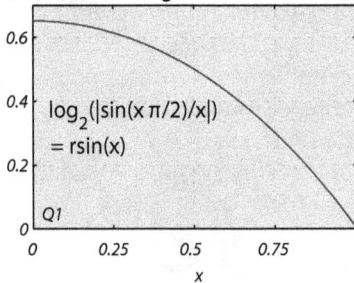

d) log2, linear domain

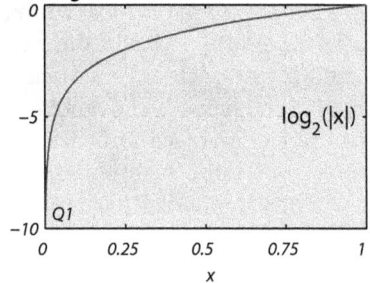

FIGURE 6.4: The first quadrants (Q1-4) of the sine function plotted in the LNS domain (a) and in the semi-logarithmic domain (b). The sine function is analytically split into (c) and (d) which can be implemented more efficiently using the LNU architecture.

SFEs. Note that these SFEs do not require any additional special case handling and range reductions in software, since this is automatically performed in HW in the LNU.

Sine and Cosine:

First, the argument range is reduced via an LNS division in the *MulDiv Block*, followed by a transformation into the linear domain by reusing the EXP functionality. The modulo-1 operation (rem) and symmetric folding can then be performed by truncating the MSBs and using integer additions, making it possible to implement the sine and cosine functions by tabulation of the first sine quadrant. Second, the evaluation of this first sine quadrant is analytically split into two terms in order to circumvent the singularity in a similar way as in the cotransformation (6.10). As shown in Figure 6.4b, the sine function $\log_2(|\sin(\pi x/2)|)$ still has a singularity in the semi-logarithmic domain. To implement it we use the decomposition

$$\log_2(|\sin(\pi x/2)|) = \log_2\left(\left|\frac{\sin(\pi x/2)}{x}\right|\right) + \log_2(|x|)$$

$$= \mathrm{rsin}\,(x) + \log_2(|x|),$$

(6.12)

since the first term (Figure 6.4c) can be efficiently approximated with a 2nd order polynomial, and the second term (Figure 6.4d) is already available as part of the *LogExp Block* in the LNU. To maximize datapath reuse these two steps are mapped onto different instructions termed SCA (sine and cosine argument), SIN and COS (the actual interpolation) which have to be issued sequentially (e.g., SIN(SCA(A,B)) with A=X, B= $\pi/2$). The SCA(A,B) is a DEX variant leveraging a cheap LNS division to divide the sine or cosine argument by $\pi/2$.

Arctangent:

The choice of the approximation scheme depends on whether only $\mathrm{atan}(x)$ or the two-argument version $\mathrm{atan2}(y,x)$ is implemented. We target the latter version in this work, such that no additional software wrapping of the arctangent function is required to calculate the correct phase. As shown in [216], the $\mathrm{atan2}(y,x)$ function can be implemented by only tabulating the $\mathrm{atan}(y/x)$ function on $y/x \in [0,1]$ (i.e., on the first Cartesian octant when interpreting x and y as coordinates). Since atan2 is an odd function, the inputs can be reduced to the first quadrant, and using the relation $\mathrm{atan}\,(y/x) = \pi/2 - \mathrm{atan}\,(x/y)$ they can be further reduced to the first octant.

a) arctan(x), double-logarithmic domain

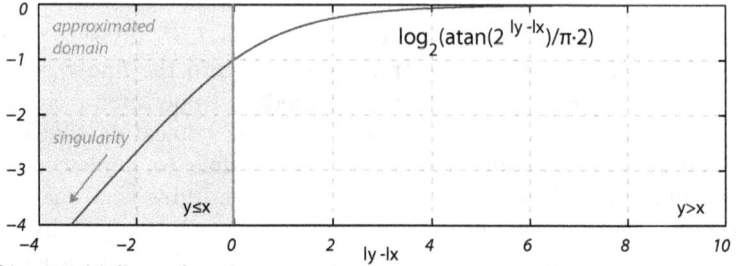

b) arctan(x), linear domain

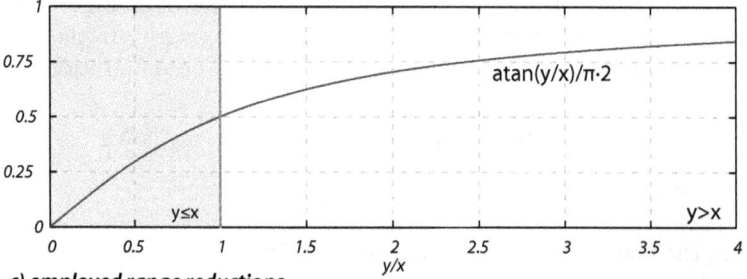

c) employed range reductions

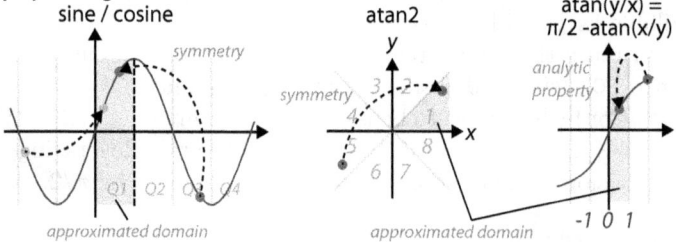

FIGURE 6.5: a) shows the arctangent function in the LNS domain, and b) in the linear domain. c) shows examples for employed range reductions.

To address the LNS singularity issue, the existing EXP and LOG functionality of the LNU is reused to approximate the arctangent in the linear domain (Figure 6.5b). The evaluation of one atan2 intrinsic is therefore split in three LNU instructions ATA (arctan argument), ATN (arctangent table lookup) and ATL (arctangent logarithm), and the corresponding evaluation sequence is ATL(ATN(ATA(A,B))), where A=Y and B=X.

The ATA(A,B) instruction is a variant of the DEX instruction. The difference is that before actually applying the EXP operation, the control logic detects cases where the result $u = y/x > 1$ and inverts the value e_u such that the result of the ATA instruction is always within [0,1]. To mirror the output of the arctangent to the correct octant, the information is encoded into the result of the ATA operation. The ATN instruction performs the actual arctangent interpolation and mirrors the result to the correct octant via the constant multiplexer in the *LogExp Block*. The ATL instruction is a LOG variant that just bypasses negative signs in this case.

Other Trigonometric Functions

Other trigonometric functions such as tan, cot, asin and acos can be easily assembled using trigonometric relations:

- *Tangent and Cotangent*: the relation

$$\tan(x) = \frac{\sin(x)}{\cos(x)} \tag{6.13}$$

  can be calculated using the four instructions SCA, SIN, COS and one division (the same holds for the cot).

- *Arcsine and Arccosine*: the relation

$$\asin(x) = \atan\left(\frac{x}{\sqrt{1.0 - x^2}}\right)$$
$$= \atan2\left(x, \sqrt{1.0 - x^2}\right) \tag{6.14}$$

  can be calculated using the six instructions MUL, SUB, SQRT, ATA, ATN, ATL (the same holds for acos).

Note that operations such as multiplications, divisions or square-roots can be executed accurately and with low-latency in the integer ALUs of the processor cores.

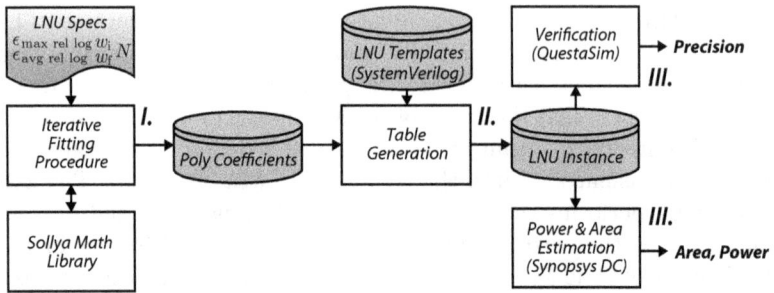

FIGURE 6.6: Overview of the LNU generation flow.

6.5   LNU Generator

The architecture presented in the previous section serves as a parameterizable template for our LNU generator. The generation flow is illustrated in Figure 6.6 and consists of three steps:

I. The generator calculates the quantized polynomial coefficients for all required functions according to the LNU specification which consists of the LNS format w_{int}, w_{frac}, polynomial order N, and error bounds $|\epsilon|_{\mathrm{max}}$, $|\epsilon|_{\mathrm{avg}}$ for the relative errors in the LNS domain.

II. All coefficients are scanned and the word width parameters for the shared datapaths are calculated. The parameters and coefficients are then printed into the architecture template in order to form a specific LNU instance.

III. The LNU instance is then exhaustively verified using RTL simulations in MENTOR QuestaSim.

The core generator functionality in steps I. and II. have been implemented in MATLAB.

6.5.1   Polynomial Fitting

For all function approximations, we use Nth order piecewise *minimax* polynomials. The coefficients are obtained using an efficient, quantization aware implementation of Remez's algorithm [207] available in the Sollya math library [221]. Since the functions used here are increasingly difficult to approximate as $r \to 0$, we subdivide their domains into logarithmically spaced segments, and within each segment a different number of piecewise polynomials is used. These segment boundaries are always aligned to multiples of powers of two, and the spacing between piecewise polynomials Δ_p^r is always a power of two. A table lookup can then conveniently be carried out by looking at the amount of leading zeros of r in order to activate a specific segment. Depending on the spacing Δ_p^r of the piecewise polynomials within that segment, the bits $[w_{\mathrm{exp}} : w_{\mathrm{frac}} + \log_2(\Delta_p^r)]$ of r are used to determine which set of polynomial coefficients to use, and the remaining LSBs are used as input into the Horner interpolator $\delta_p^r = r[w_{\mathrm{frac}} + \log_2(\Delta_p^r) - 1 : 0]$.

The fitting procedure subdivides the function to be fitted into logarithmically spaced segment (e.g. $[0, 32)$ is split into $[16, 32)$, $[8, 16)$, $[4, 8)$, etc.), and on each segment it fits a set of piecewise polynomials with an initial spacing $\Delta_p^r = 0.25$. This choice has been made to implicitly limit the integer word widths in the shared interpolators. If the error of this piecewise polynomial is too large, the spacing is iteratively divided by 2 until the error requirements can be fulfilled. Once all segments have been processed, adjacent segments with equal spacing are grouped together in order to facilitate table lookup, and the coefficients are handed over to the table generation.

6.5.2 Error Calculation and Word Width Selection

To asses the overall error of a given piecewise polynomial, we exhaustively assess it at all bit combinations of r within its domain and compare the result to a double precision reference. The errors so calculated contain both the approximation error stemming from the polynomial approximation, and the quantization errors from the coefficient quantization and intermediate truncation/rounding steps. The impact of coefficient quantization is minimized by using a quantization aware minimax fitting method. Since a rounding step always incurs a carry propagation in hardware, we only perform one rounding step to the output precision w_{frac} at the end of a polynomial evaluation. Intermediate results after multiplications are always truncated to the fraction of the next coefficient in the Horner scheme, as illustrated in Figure 6.3. The word widths of the polynomial coefficients are determined using a similar heuristic as described in [222]. I.e., since typically $\delta_p^r < \Delta_p^r < 1$, the weight of LSBs of a polynomial coefficient p_i^r are decreased by at least a factor of Δ_p^r in each multiplication with δ_p^r. Therefore, we heuristically determine the fraction width $w^r_{\text{frac } p_i}$ of higher order coefficients as

$$w^r_{\text{frac } p_i} = \max\left(0, w^r_{\text{frac } p_0} + i \cdot \log_2\left(\Delta_p^r\right)\right) \qquad (6.15)$$

and the fraction of the 0th order coefficient is set to

$$w^r_{\text{frac } p_0} = w_{\text{frac}} + \max\left(0, n_{\text{guard}} - \lceil\log_2\left(\epsilon_{\text{max rel log}} - \epsilon_{\text{rnd}}\right)\rceil\right) \qquad (6.16)$$

where $\epsilon_{\text{max rel log}}$ is the maximum error requirement, and the term $\epsilon_{\text{rnd}} = 0.5$ amounts for the maximum error due to the final rounding step. The amount of additional guard bits has been set to $n_{\text{guard}} = 3$, since fewer than 3 bits lead to a significant increase in table size.

Within the critical region, where two polynomial results $p(r)$ and $q(r)$ are added up before the final rounding step, the error requirement for both polynomials is adjusted to $\tilde{\epsilon}_{\text{max rel log}} = (\epsilon_{\text{max rel log}} - \epsilon_{\text{rnd}})/2$ in order to account for the fact that we have twice as many error sources now. Error checking of the individual polynomials is performed without final rounding in this case to make sure that the polynomials are precise enough before being added. Also, ϵ_{rnd} is set to 0 in (6.16), as it has already been accounted for in $\tilde{\epsilon}_{\text{max rel log}}$.

6.6   Processor Integration

To evaluate the performance of different LNUs in a shared setting, we have designed[4]a multi-core processor system based on a 32 bit OpenRISC core [223] using the UMC 65 nm LL technology. As shown in Figure 6.7a, the system consists of four cores with 1 kB private instruction caches that share a single LNU and contains a total of 32 kBytes of memory.   The four-to-one sharing ratio is motivated

---

4. Note that the LNU project has been carried out jointly with Michael Gautschi (see also [38–40]). While he concentrated on the integration of the LNU into the PULP multi-core processing system, the cluster-level simulations and the chip backends, I have concentrated on the LNU architecture, SFEs and the LNU generator. While the technical work can be clearly attributed to either Michael Gautschi or me, the

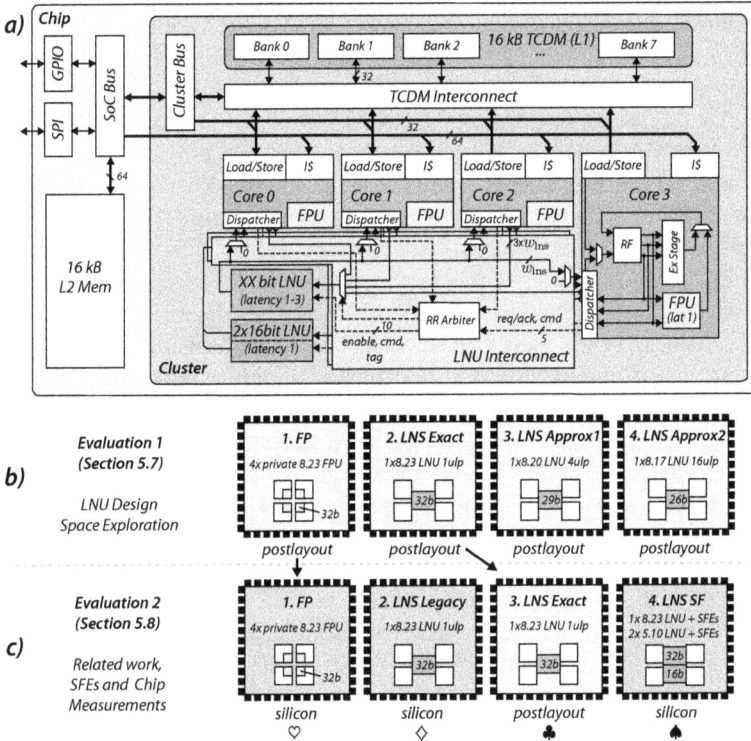

FIGURE 6.7: a) Integration of the shared LNUs and the 16 bit vector LNU into an OpenRISC cluster. For clarity, connection details of the LNU are only shown for core 3. b) and c) show the designed chip variants for the two evaluations in Sections 6.7 and 6.8. The symbols ♡♢♣♠ are used as short IDs in all tables and figures in Section 6.8.

by the fact that in most FP programs, the fraction of ADD/SUB instructions, rarely exceeds 0.25 (see Figure 6.10a for examples). For comparison, an identical system has been designed featuring four cores with private IEEE 754 32 bit compliant FPUs including HW support for ADD, SUB, MUL and typecasts. For DIV we use SW emulations as described in Section 6.7.2 which is a common approach of adding FP support to small embedded processors [209]. The implemented FPU is

---

initial idea of sharing the LNU in a multi-processor system, as well as the obtained energy-efficiency results and function kernel performance analyses are difficult to disentangle and thus represent joint contributions. They have been included in the forthcoming sections for completeness.

a shared normalizer design similar to [208] (but without divider) with a complexity of ~11 kGE, which is competitive with state-of-the-art implementations [209, 210, 224].

Two different evaluations are carried out in Sections 6.7 and 6.8. The first evaluation concentrates on tradeoffs between exact and approximate LNU instances, and the second evaluation on related work, the SFEs and chip measurements. In both evaluations we use the multi-core processor system designed in slightly different configurations. In Section 6.7, we use it to compare two approximate shared-LNU designs (*LNS Approx1* and *LNS Approx2*) with an exact shared-LNU design (*LNS Exact*) and with the FPU baseline (*FP Exact*). These four designs are listed in Figure 6.7b, and all generated results are from postlayout simulations. The specific LNU parameterisations will be explained in more detail in Section 6.7. In Section 6.8, we compare the four designs shown in Figure 6.7c, where three of the designs have been taped out and measured. The last chip is extended with a vectorized half-precision LNU allowing to perform two half-precision LNU operations in parallel as well as providing dot-product (DOTP) instructions that utilize the *MulDiv Blocks* of both half-precision LNUs. All clusters have been designed to run at 500 MHz at 1.2 V under typical PTV conditions. To meet the timing constraints of the cluster, the FPU and the 16 bit vector LNU have been pipelined once and the other LNUs one to three times, depending on the depth of the combinational logic.

Note that the tool flow has evolved over the project and the newest designs (*LNS Approx1, LNS Approx2, LNS Exact, LNS SF*) are more efficient w.r.t. the first silicon implementations of the *FP Exact* and *LNS Legacy* designs [38]. Therefore, the backends have been repeated using the current design flow to compensate for any systematic offsets in the measurements. The cotransformation has also evolved over the project, and while all new designs use [182] described in Section 6.3.4, only the silicon implementation named *LNS Legacy* used an LNU described in detail in our previous work [38, 53]. The main difference between the *LNS Legacy* and the other designs (*LNS Approx1, LNS Approx2, LNS Exact, LNS SF*) is the way how subtractions in the critical region are handled. The new designs presented here have a smaller interpolator and require about 45% less LUT bits than the *LNS Legacy* design (see Table 6.3).

6.7 LNU Design Space Exploration

In this section we first present a design space exploration of a family of exact, faithful and approximate LNUs (without SFEs) generated with our framework. The resulting instances have been synthesized using the 8-metal UMC 65 nm LL CMOS technology with SYNOPSYS Design Compiler version 2015.6 in order to get gate-level area and timing estimates at typical conditions, 25 °C and 1.2 V supply. For cluster level evaluations, we have selected a set of representative LNU versions, pipelined them using the automatic retiming feature of SYNOPSYS Design compiler, and for each cluster version we have performed a complete back-end design flow using CADENCE EDI 14.24 in the same 8-metal UMC 65 nm LL CMOS technology. We have modified the back-end of the OpenRISC LLVM compiler to support the LNS format, and added new instructions to support the additional functionality provided by our LNU. A set of benchmarks written in C was compiled and executed on the FP and LNU cluster architectures, which have been simulated in MENTOR QuestaSim 10.3a using back-annotated post-layout gate-level netlists. Finally, the value change dump (VCD) files obtained were used to analyze the power dissipation in CADENCE EDI 14.24.

6.7.1 Area, Precision and Timing Tradeoffs

There are three approaches to trade-off circuit area, delay and precision: changing the word width of the design, changing the interpolation order, or relaxing the precision requirements. In order to achieve smaller and faster designs, usually the word width and interpolation order are reduced and tables are calculated for exact or faithful representations. In this paper, for a given word width we also relax the precision requirements up to 16 ULP which allows simpler tables that reduce the overall circuit size. In Table 6.1 we show the normalized area of 1st and 1st order LNUs under different word widths and approximation goals. As we can see, relaxing the precision of an exact 2nd order LNU from 0.72 ULP to 8 ULP leads to an area reduction of 40%. The interesting result from Table 6.1 is that area-wise, similar results can be obtained by either reducing the word width or the precision. However, when considering the average error,

the situation changes. In Figure 6.8 the *average* error for LNU designs
are plotted against the circuit area. In this plot we can see that
approximate configurations with larger word width are consistently
more accurate (on average) than lower word width configurations.
E.g., a 2nd order 8.20 with 8 ULP precision LNU configuration is not
only slightly smaller than a 2nd order 8.17 with 0.72 ULP precision,
but the average error is lower by a factor of ~2.74. Another observation
that can be made from Table 6.1 is that for higher precision ranges
(17-23 fraction bits), 2nd order LNUs are much more area efficient as
fewer LUT entries are required. For designs with a $w_{\mathrm{frac}} \leq 14$, 1st
order interpolation is preferable.

Reducing circuit complexity has an additional benefit as it also
reduces the critical path through the LNU. Depending on the clock
frequency of the system where the LNU will be integrated, this could
change the number of required pipeline stages which in turn can have
important consequences on the overall performance of the system.
Figure 6.8 shows a design space exploration for 50 LNU configurations, mapped to hardware with different timing constraints. The target
clock period (in our case 2 ns for 500 MHz operation) is overlayed in
this graph, and it can be seen that LNUs with different area/precision
trade-offs can be obtained with 2 to 4 pipeline stages. We have selected
a representative set of three LNU variants

- *LNS Approx2* (8.17 bit, 16 ULP, 1st order),
- *LNS Approx1* (8.20 bit, 4 ULP, 2nd order),
- *LNS Exact* (8.23 bit, 0.72 ULP, 2nd order),

that were implemented with different numbers of pipeline stages to be
evaluated in the next section comparing overall system performance.

6.7.2   Function Kernel Performance

After evaluating the performance of single LNUs, we now present a
more detailed performance analysis of a shared LNU in a real multi-
core system running actual computation kernels. For this comparison
we use a system comprised of four 32 bit OpenRISC processor cores
running at 500 MHz in the UMC 65 LL technology used throughout this work. Our reference (*FP Exact*) is a system that includes
four IEEE 754 single-precision compliant FPU units with support for

TABLE 6.1: Relative area comparison (in percent) of exact and approximate LNUs synthesized with 4.5 ns timing constraint.

Interp. Order	$w_{int} . w_{frac}$	Precision Constraint in LNS Domain [ULP]					
		0.72	1	2	4	8	16
	8.23	618.6	232.7	153.6	116.0	86.6	73.0
	8.20	218.6	106.5	72.4	53.8	46.7	39.6
$N = 1$	8.17	89.8	48.2	38.0	31.0	26.9	25.4
	8.14	36.3	26.8	22.3	19.8	18.3	17.2
	5.10	15.5	12.2	11.8	11.0	10.3	9.4
	8.23	100.0	76.5	68.4	62.9	59.9	57.8
	8.20	69.5	56.5	51.2	48.2	46.1	44.4
$N = 2$	8.17	49.0	41.3	38.5	37.0	36.4	34.9
	8.14	34.9	31.1	29.4	28.3	27.0	24.9
	5.10	21.4	18.2	16.9	15.9	14.6	13.3

ADD/SUB/MUL and casts. This is compared against three different LNU configurations (*LNS Exact*, *LNS Approx1*, and *LNS Approx2*) selected from the design space exploration described previously. Table 6.2 lists all four compared variants and their complexities. As can be seen, all LNU variants are smaller than the reference FPU implementation. The larger size of the LNUs is compensated by sharing them among the processor cores. Note that, even though these systems only have a single LNU, they can perform up to four MUL/DIV/SQRT single-cycle LNS operations within the integer ALU of the cores.

The reference FPU at 11 kGE is compact but does not include support for more complex operations which have to be emulated in software. For DIV operations, we perform a range reduction to [1,2) and generate a linear estimate for the inverse that is refined using three Newton-Raphson iterations. A similar technique is used for the SQRT, where the initial estimate is generated using the fast-inverse square-root [225].

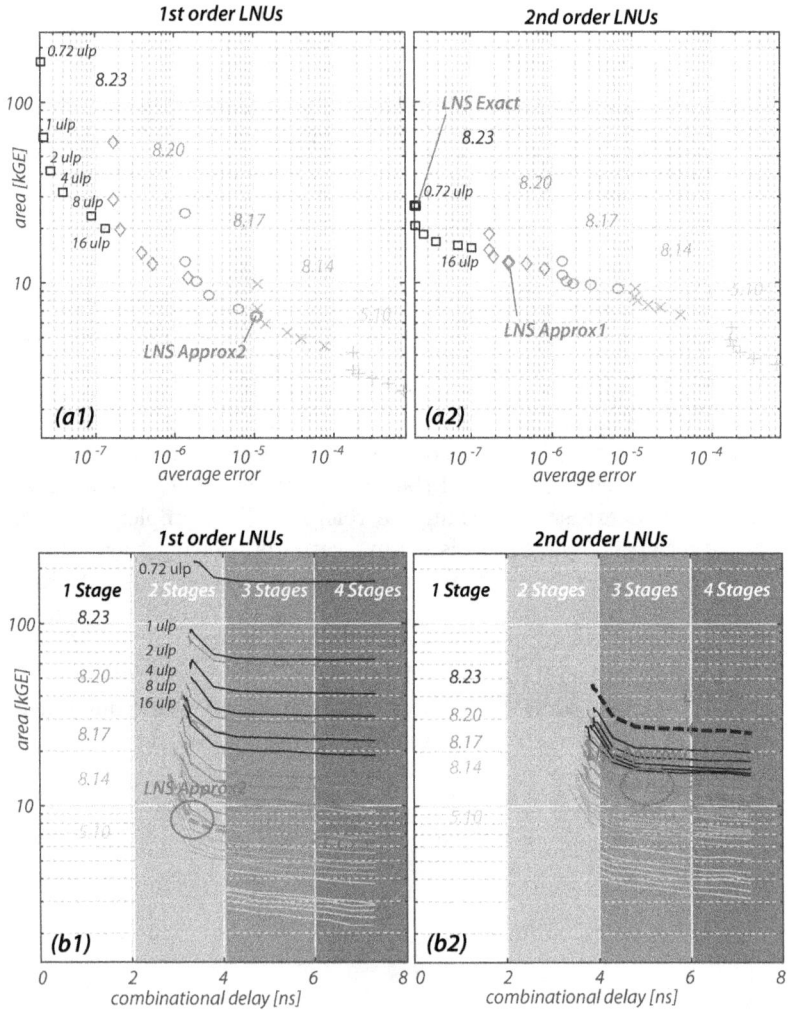

FIGURE 6.8: a) Area vs. average error (in the FP domain) for LNUs synthesized with 4.5 ns timing constraint. b) area time (AT) plots, including 0.8 ns I/O delay for the sharing infrastructure.

EXP/LOG operations combine range-reduction with standard, high-order interpolation as described in [217]. Note that this is a common way to add support for these functions [209]. Table 6.2 lists the number of cycles per instruction and the corresponding energy consumption in pJ/Op. While ADD/SUB operations are (as expected) costlier in the LNUs than in the FPU, all other operations can be performed more efficiently.

TABLE 6.2: Instruction latency and energy efficiency of the FPU and LNU clusters.

Format	IEEE 754	LNS		
Name	FP Exact	LNS Exact	LNS Approx1	LNS Approx2
Word Width	8.23	8.23	8.20	8.17
Precision	0.5 ULP	0.72 ULP †	4 ULP †	16 ULP †
Order N	-	2	2	1
FPU/LNU	4×11 kGE	36 kGE	27 kGE	23 kGE
Total Area	720 kGE	718 kGE	708 kGE	704 kGE
Operations	Latency [cycles] / Energy at 1.2 V [pJ/Op]			
I2F/F2I	2 / n.a.	4 / n.a.	3 / n.a.	2 / n.a.
ADD	2 / **41**	4 / 106	3 / 89	2 / 85
SUB	2 / **40**	4 / 110	3 / 92	2 / 90
MUL	2 / 48	1 / 31	1 / **28**	1 / 31
DIV	62 / 525*	1 / 32	1 / **29**	1 / 32
SQRT	56 / 609*	1 / 16	1 / **14**	1 / 15
EXP	51 / 567*	4 / 115	3 / **57**	2 / 87
LOG	85 / 696*	4 / 105	3 / **55**	2 / 74

* software emulation. † in the LNS domain.

We have compiled a set of benchmarks written in C to reflect a variety of different signal processing applications. The benchmark set consists of basic linear algebra subprograms (BLAS) like AXPY, GEMM and GEMV, geometry calculations (2D homographies, reprojection error [28], 3D distances), matrix decompositions (QR, CHOL), regression (radial basis functions), and image processing (bilateral filter (BF) and FIR filters, gradient magnitude, DCT) and audio processing kernels (sine generation, Butterworth IIR lowpass). The FP instruction ratio and the instruction mix of the benchmark applications are shown in Figure 6.10a. As can be seen, the ratio of ADD/SUB operations for most benchmarks is below 25% further reinforcing our sharing concept.

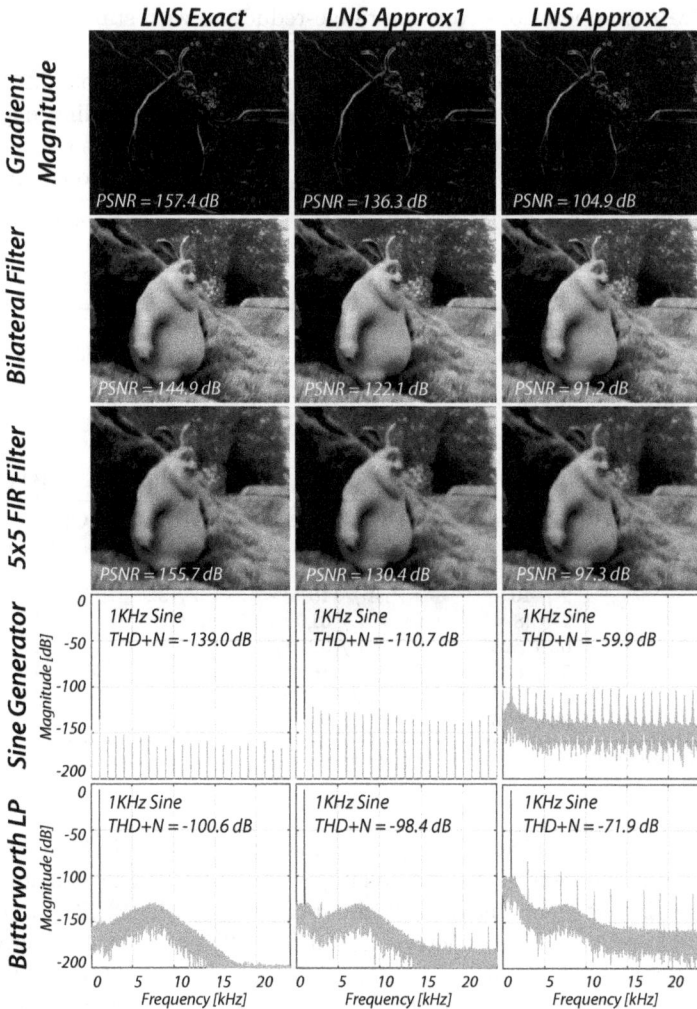

FIGURE 6.9: Images and audio streams processed with the LNS cluster variants. Images copyright © 2008, BLENDER FOUNDATION, www.bigbuckbunny.org.

All performance evaluations were made with four configurations of the cluster, including two configurations (*LNS Approx1* and *LNS Approx2*) that have slightly relaxed precision. We have determined the impact of this range of approximations on image and audio processing kernels using PSNR and total harmonic distortion plus noise (THD+N) metrics which are shown in Figure 6.9. The *FP Exact* and *LNS Exact* configurations deliver identical results. We can observe that for all image processing kernels the PSNR values stay way above the 30 dB mark, below which artifacts start to be visible. Also for the audio processing kernels we see that THD+N values are below -59 dB for all LNUs. For all practical purposes, both approximate versions show no perceptible quality degradations. Normalized execution time and energy efficiency improvements with respect to the reference implementation (*FP Exact*) mirror each other and are shown in Figure 6.10c and d. In the best case, a speedup of 5.54× can be achieved for the DIST3D case using *LNS Approx2*. Even when the *LNS Exact* configuration is used, on average the kernels can be calculated 1.71× faster and 1.65× more efficiently in terms of energy. As expected the shared LNU has the most difficulty with kernels containing many ADD/SUB operations (such as the linear algebra or FIR/IIR filters). Even so, it can be seen in Figure 6.10c and d that the *LNS Approx2* variant can perform at least as well as the FPU design on all kernels except the Butterworth filter. We also observe that the pipeline depth of the LNU has significant impact on the overall system performance. In our example, the 4-stage in-order OpenRISC cores can be operated without stalling for LNUs with 2 stages. LNUs with more stages incur additional stalls, reducing the overall instructions per cycle (IPC) as seen in Figure 6.10b.

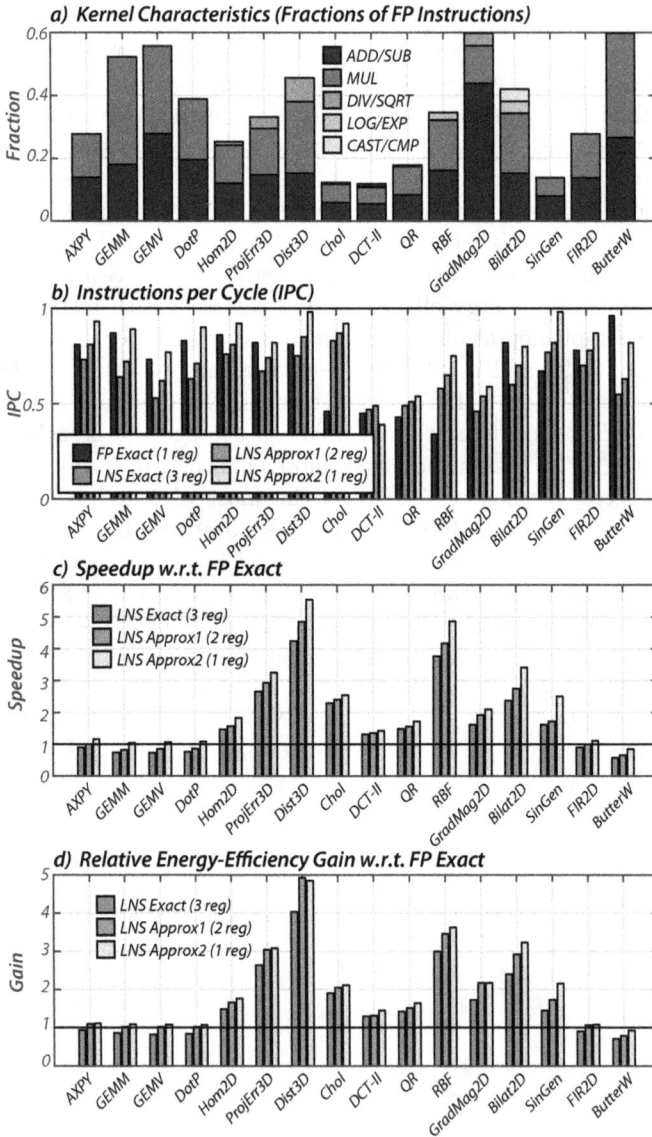

FIGURE 6.10: Application level performance for the four cluster variants.

6.8 LNUs in a Larger Context

In this section, we performed similar evaluations as in the previous Section 6.7 by integrating generated LNUs into an OpenRISC cluster as described in Section 6.6. The main difference to the previous evaluation are the chosen LNU configurations (exact LNUs, now also including SFEs), as well as the newer tool versions. We used SYNOPSYS Design Compiler 2016.03 for synthesis, and the back-end flows have been done using CADENCE Innovus-15.20.100 in the same 8-metal UMC 65 nm LL CMOS technology. Intrinsics and vector instructions are provided as LLVM compiler built-in functions (compiler auto-vectorization is currently not supported). A set of benchmarks written in C was compiled and executed on the FP and LNS architectures. The *FP Exact, LNS Legacy* and *LNS SF* designs (see Figure 6.7c) have been manufactured, extensively tested and measured. The remaining *LNS Exact* cluster has been simulated in MENTOR QuestaSim 10.3a using back-annotated post-layout gate-level netlists and power has been analyzed in CADENCE Innovus-15.20.100. First, a comparison of our generated LNU designs with related LNUs and SFUs is given in Section 6.8.1, followed by a comparison of the designed chips in Section 6.8.2. Finally, we will give detailed results of instruction- and kernel-level efficiencies in Section 6.8.3 and Section 6.8.4.

6.8.1 Comparison with Related Work

The AT behavior of the compared LNU configurations and a detailed area split for an LNU with SFEs is shown in Figure 6.11a and b. A large part of the LNU area is occupied by LUTs (42.3%) and interpolators (42.8%). Table 6.3 shows a comparison of synthesis results for combinational implementations in both 65 nm and 180 nm. Our new LNU without SFEs is significantly smaller in terms of normalized gate-equivalents (GE) than most related designs, and provides more functionality such as casts and EXP/LOG functions. The additional FMX/TRIG instructions increase the area by only 17%. Compared to the best related 65 nm design *Minimax (2CT)* by [189] our new LNU has similar area, but a higher combinational delay. Note however, the only silicon-proven related design is the ELM LNU [187, 188] and *Minimax (2CT)* has first been developed in 180 nm, and the 65 nm results listed in Table 6.3 represent additional technology translations

a) AT Behaviour in 65nm

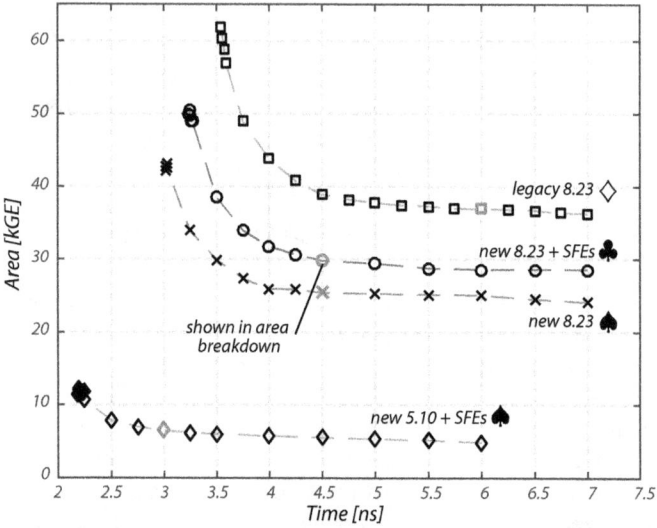

b) Detailed Area Breakdown

new 8.23+ SFEs ♠
29.8 kGE @ 4.5ns

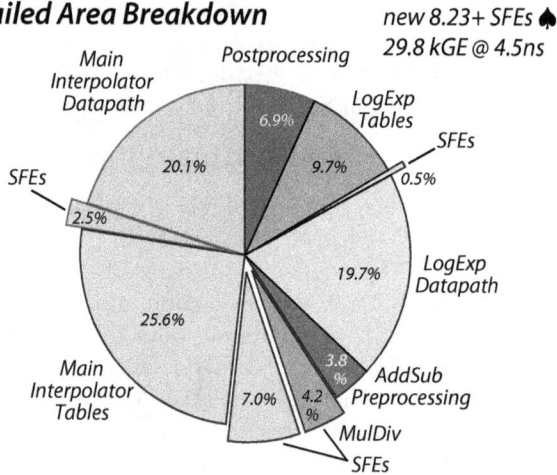

FIGURE 6.11: a) Post-synthesis AT behaviours of the LNUs analyzed in this work (combinational, unpipelined entities). The LNUs listed in Table 6.3 correspond to the red AT design points in this figure. b) detailed area breakdown of the new LNU with SFEs (overheads due to these SFEs are exploded in the pie chart).

provided in [189]. The related designs [187–189] either use custom 1st order schemes, or 2nd order polynomial interpolators with a dedicated squarer. These schemes use 1.8-5.9× more LUT entries than our design, but have the advantage of providing lower latency compared to the 2nd order Horner interpolator used in our LNU. However, these designs exhibit variable latency, since critical region subtractions have to pass through the interpolator twice, making it harder to share such an LNU. A post-layout comparison of the new LNU including SFEs with an SFU from NVIDIA Tesla GPUs [192,212] and from Caro et al. [213] is shown in Table 6.4. Our design is comparable in complexity while providing equivalent accuracy levels and functionality.

Note that an LNS-based processor does not require any additional units to be complete, whereas an FP-based system additionally requires an FPU for standard operations. When comparing the LUTs, we can observe that the design by Caro et al. [213] requires the least amount of bits for special functions (SFs). The reason for this is that they additionally enforce constraints among neighboring LUT segments when fitting the non-linear functions. This allows to obtain even smaller LUTs than with the minimax fitting heuristics employed in [192, 212] and our design. However, the design by Caro et al. does not have support for LNS ADD/SUB, trigonometric functions and the special cases (like NAN/INF). Note that the complexity of the design by [192, 212] is specified in full-adder equivalents (FE) of a proprietary library and has been converted assuming 5.25 GE per FE. Also, the trigonometric intrinsics for sine and cosine do not support range reduction from arbitrary values to the first function period. Our design in contrast is able to perform automatic range reduction on all trigonometric functions thanks to cheap LNS divisions.

TABLE 6.3: Comparison of different LNUs and related designs (synthesis results of combinational, unpipelined entities). 1GE in UMC 180 nm corresponds to 9.374 μm², and 1GE in UMC 65 nm corresponds to 1.44 μm².

	[187] ELM	[188] ModChes2	[189] Minimax2	Legacy	This Work New	This Work New+SFEs	This Work New+SFEs
Functionality	ADD, SUB			ADD, SUB, F2I, LOG, EXP	ADD, SUB, I2F, EXP	ADD, SUB, I2F, LOG, EXP, FMX, TRIG	ADD, SUB, I2F, LOG, TRIG
Interpolation							
Wordwidth [bit]	32	32	32	32	32	32	16
cotr Type	[185]	[188]	[188]	[53]	[182]	[182]	[182]
cotr Interp. N	custom 1	custom 1	minimax 2	custom 1	minimax 2	minimax 2	minimax 1
Other Funcs N	custom 1	custom 1	minimax 2	minimax 2	minimax 2	minimax 2	minimax 1
Error§ [ULP]							
ADD (max)	0.4544	0.4623	0.4944	0.4618	0.3806	0.3806	0.3753
ADD (avg)	0.1777	0.1745	0.1721	0.1748	0.1744	0.1744	0.1734
SUB (max)	0.4952	0.4987	0.4626	0.4786	0.4755	0.4755	0.4561
SUB (avg)	0.1776	0.1738	0.1719	0.1748	0.1750	0.1750	0.1738
LUT Size [kBit]							
F^+ / F^-	227	162	?	20 / 41	30 / 12	30.4 / 12.2	2.8 / 0.6
cotr / other	129 / –	21 / –	? / –	24 / 28	5 / 13	5 / 24	1.0 / 2.6
total	356	183	110	113	61	71.5	7.0
UMC 180 nm							
Del. min/max [ns]	11.7/13.5	7.1/14.8	9.3/~19.2	17.0/17.0	12.5/12.5	12.5/12.5	8.5/8.5
Area [mm²]	0.91	0.59	0.47	0.41	0.30	0.38	0.08
Area [kGE]	97	63	51	44	32	40	8
Silicon Proven	yes (ELM)	no	no	no	no	no	no
UMC 65 nm							
Del. min/max [ns]	–	0.9/1.9†	1.2/2.6	6.0/6.0	4.5/4.5	4.5/4.5	3.0/3.0
Area [mm²]	–	0.04†	0.03	0.05	0.04	0.04	0.01
Area [kGE]	–	27†	22	38	25	30	7
Silicon Proven	–	no	no	yes	no	yes	yes
Used in Design LNS .	–	–	–	(Legacy ◇)	(Exact ♣)	(SF ♠)	(SF ♠)

† 65 nm numbers from [189].　§ relative errors in the FP domain ($|\epsilon|$max rel arith).

TABLE 6.4: Comparison of the new LNU with SFEs and FP SFUs (post-layout figures, i.e. pipelined, placed&routed units).

Function	NVidia [192, 212]				Caro et al. [213]				This Work (New+SFEs ♠)			
	Table Range	Error×	LUT [kBit]	Lat/ TPut	Table Range	Error×	LUT [kBit]	Lat/ TPut	Table Range	Error×	LUT [kBit]	Lat/ TPut
$\sqrt{x}$	-	-	-	-	[1,2]	1.0 ULP	1.73	6/1	-	0.69 ULP	-	1/1§
$\sqrt{2x}$	-	-	-	-	[1,2]	1.0 ULP	1.82	6/1	-	0.69 ULP	-	1/0.5§
$1/\sqrt{x}$	[1,4]	1.52 ULP	6.5	?/1	[1,2]	1.0 ULP	3.71	6/1	-	0.69 ULP	-	1/0.5§
$1/\sqrt{2x}$	-	-	-	-	[1,2]	1.0 ULP	3.65	6/1	-	0.69 ULP	-	1/0.33§
$1/x$	[1,2]	0.98 ULP	6.5	?/1	[1,2]	1.0 ULP	4.29	6/1	-	0 ULP	-	1/1§
$\log_2(x)$	[1,2]	22.57 bit	3.25	?/1	[1,2]	24 bit	4.03	6/1	[1,4]	0.75 ULP	9.15	4/1
2^x	[0,1]	1.41 ULP	3.25	?/1	[0,1]	1.0 ULP	1.98	6/1	[0,1]	23.0 bit	4.48	4/1
$\sin(x), \cos(x)$	[0,π/2]	1.72e-7	3.25	?/1	-	-	-	-	[0,1]	1.85e-7	6.54	8/0.5
$\operatorname{atan2}(y,x)$	-	-	-	-	-	-	-	-	[0,1]	1.87e-7	4.03	12/0.33
Other Support												
Functions	2D Attribute Interpolation				no				ADD, SUB, I2F, F2I, FMA			
NaN, INF	?				no				yes			
Implementation												
Technology	?				TSMC 180 nm				UMC 65 nm			
1GE [μm²]	?				9.374†				1.44			
Freq. [MHz]	?				420				305			
Area [mm²]	?				0.34				0.07			
Area [kGE]	8030 FE ≈ 42 kGE ‡				36				45			

† Assumed GE for 180 nm.   ‡ Assuming 5.25 GE per FE.   § Computed in integer ALU of the processor cores.
× Error either specified as $|\epsilon|_{\text{max rel arith}}$ or as $|\epsilon|_{\text{max abs arith}}$ if not applicable. For $\log_2(x)$ (in FP) and 2^x (in LNS) precision given in amount of good bits returned, see [212, 213].

6.8.2   Chip Variants Designed

Table 6.5 lists all designs and gives a comparison of the measured and simulated chip variants. In terms of related work, there exist many application-specific implementations in the literature (e.g., [176, 197–200, 214]) which use LNS and its benefits to compute faster and/or more energy-efficiently. However, there are surprisingly few designs where an LNU equivalent to single-precision FP is designed and integrated into a programmable processor. In fact, the only comparable ASIC design where this has been done is the ELM processor [187], which is also listed in Table 6.5 for comparison.

6.8.3   Instruction Level Performance

Figure 6.12a shows an efficiency trade-off analysis similar to the one conducted in [210], enhanced with datapoints from related FPU designs [210, 224]. Figure 6.12b shows the energy efficiencies over a range of different VDD conditions for a selection of operations. A complete comparison of the operator efficiencies of all designs is given in Table 6.6. While LNS ADD/SUB are clearly less energy-efficient than the FP equivalents, LNS MUL requires ~30% less energy than in FP for all LNS designs. Apart from these basic instructions, LNS supports extremely energy-efficient, single-cycle square-roots (6 pJ/Op @0.8 V) and divisions (14 pJ/Op @0.8 V) utilizing the shifter and adder of the processor ALUs. Also complex functions such as EXP and LOG can be computed in the LNU for only 45 pJ/Op and 43 pJ/Op, respectively (@0.8 V). The architectural improvements of the LNU result in 27% more efficient ADD/SUB instructions when comparing the *LNS Legacy* design with *LNS Exact*. The trigonometric function extensions of *LNS SF* cause a small increase in LUT size, but at the same time enable range-reduced sine/ cosine and arctangent function instrinsics with 2-3 LNU instructions and a low energy consumption of 127-183 pJ/Op @0.8 V. In addition, the *LNS SF* design supports 16 bit instructions which have a 1.6-2.8× lower energy consumption w.r.t. their 32 bit equivalents.

TABLE 6.5: Comparison of LNU-based and FPU-based Processor Chips.

		ELM [187]	1. FP Exact ♡	2. LNS Legacy ◇	3. LNS Exact ♣	4. LNS SF ♠
Results From		Silicon	Silicon	Silicon	Post-layout	Silicon
Technology		180 nm	65 nm	65 nm	65 nm TT 25°C	65 nm
Area [kGE]		-	676	702	682	725
Area 1 Core [kGE]		-	39.6	41.3	41.4	43.4
FPU/LNU Area [kGE]		-	4×9.6	57.1	37.7	69.2
Supply Voltage [V]		-	0.8 – 1.3	0.8 – 1.3	1.2	0.8 – 1.3
Max. Freq. @1.2 V [MHz]		125	374	337	500	305
Power@100MHz, 1.2 V [mW]		-	23.2	24.4	22.2	24.5
Avg FPU/LNU Utilization		-	0.22	0.39	0.39	0.42
LNU/FPU Type		ELM LNU	FPU	Legacy LNU	New LNU	New LNU+SFEs
Wordwidth [bit]		32	32	32	32	32 2×16
Max Error§ [ULP]		0.454	0.5	0.478	0.476	0.476 0.456
Latencies						
ADD/SUB/FMA	HW	3/3(4)†/-	2/2/-	4/4/-	4/4/4	4/4/4 2/2/2
	SW	-/-/-	-/-/-	-/-/-	-/-/-	-/-/- -/-/-
MUL/DIV/SQRT	HW	1/1/1	2/-/-	1/1/1	1/1/1	1/1/1 1/1/1
	SW	-/-/-	-/62/56	-/-/-	-/-/-	-/-/- -/-/-
EXP2/LOG2/casts	HW	-/-/-	-/-/-	4/4/4	4/4/4	4/4/4 2/2/2
	SW	-/-/-	51/85	-/-/-	-/-/-	-/-/- -/-/-
sin/cos/atan2	HW	-/-/-	-/-/-	-/-/-	-/-/-	8/8/12 4/4/6
	SW	-/-/-	69‡/68‡/340	64/60/92	64/60/92	-/-/- -/-/-

† Variable latency LNU. ‡ Without range-reducing division (phase magnitude $\geq 2\pi$). § In the FP domain ($|\epsilon|_{\max \text{ rel arith}}$).

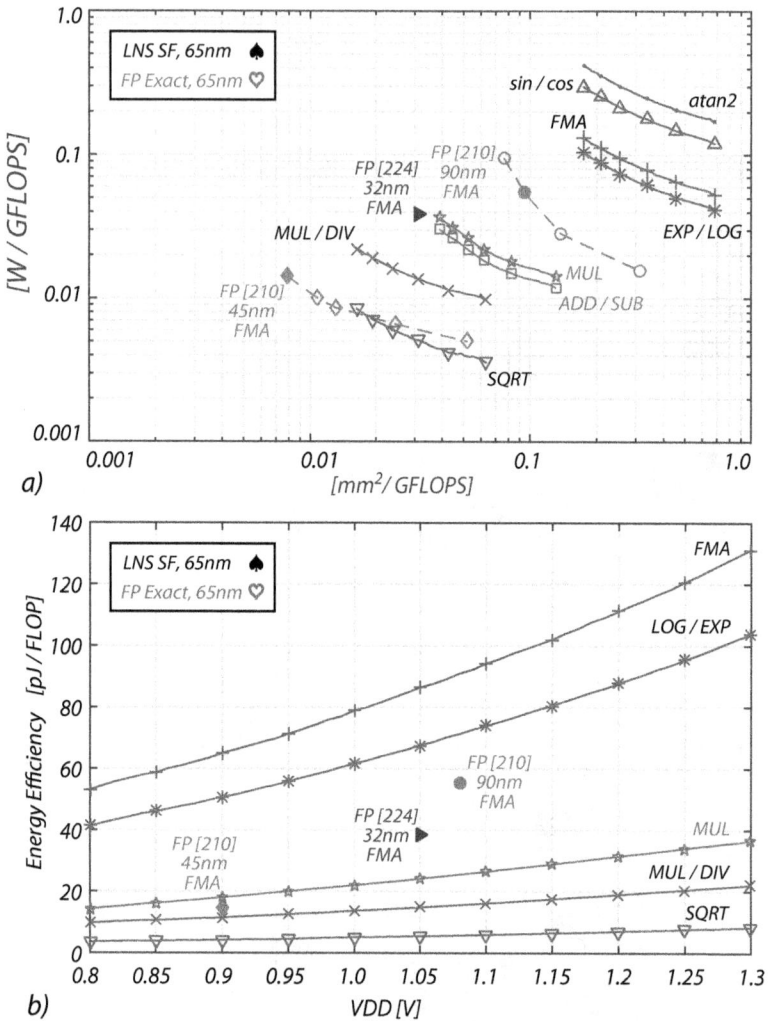

FIGURE 6.12: a) W/GFLOPS vs. sqmm/GFLOPS and b) pJ/FLOP vs. supply voltage. The curves of our designs in a) have been obtained by sweeping the supply over the range 0.8-1.3 V. Curves for related designs [210] represent differently pipelined designs. The LNS MUL/DIV/SQRT operations in a) have been normalized with the ALU area of one processor core (0.006 mm² or 4 kGE). The processor overheads mentioned in Table 6.6 have been subtracted from all shown results.

TABLE 6.6: Measured and Simulated Energy Consumption of Single Operations.

Design	Galal et al. [210] Estimations		Intel [224]	1. FP ♡	2. LNS Legacy ◇	3. LNS Exact ♣	4. LNS SF ♠	
Results From	Estimations		Silicon	Silicon§	Silicon§	Post-layout§	Silicon§	
Technology	90 nm	45 nm	32 nm	65 nm	65 nm	65 nm TT 25°C	65 nm	
Vdd [V]	1.08	0.9	1.05	0.8	0.8	0.8	0.8	
Frequency [MHz]	1200	2080	1450	165	115	150	97	
FP Width [bit]	32	32	32	32	32	32	32	2×16
Pipeline Depth	10	6	3	2	4	4	4	2
Area [mm²]	0.113	0.016	0.045	0.015	0.089	0.054	0.066	0.017
[pJ/FLOP]								
ADD	-	-	-	**16**	55	40	58	50
SUB	-	-	-	**16**	55	44	63	64
MUL	-	-	-	**18**	13	13	14	16
DIV	-	-	-	-	**13**	**13**	14	17
SQRT	-	-	-	-	**6**	**6**	**6**	8
EXP2	-	-	-	-	43	**31**	45	37
LOG2	-	-	-	-	48	**31**	43	34
FMA†	55	-	39	-	-	-	57	44
MEX	14	-	-	-	-	-	57	51
sin(x)	-	-	-	-	-	-	127	92
cos(x)	-	-	-	-	-	-	127	96
atan2(y,x)	-	-	-	-	-	-	183	152

† One 1 FMA = 1 floating point operation (FLOP) in this comparison.
§ Background power (NOP) subtracted, includes core activity (estimated core overheads are on average 2.6 pJ/FLOP for single-operand instructions and 4.2 pJ/FLOP for all other instructions).

6.8.4   Function Kernel Performance

To analyze the performance of the shared LNU in the multi-core cluster, a representative set of benchmark kernels (Table 6.7) reflecting different signal processing applications has been compiled. As in our previous evaluation in Section 6.7, the set ranges from linear algebra operations, geometry calculations, matrix decompositions and regression to image and audio processing kernels. The FP instruction ratio and the instruction mix of the benchmarks are shown in Figure 6.13a.

The reference FPU is compact but does not include support for more complex operations which have to be emulated in SW. DIV, SQRT, EXP and LOG are implemented as described in Section 6.7.2. The additional trigonometric function extensions combine periodic range rediction standard high-order interpolation [217]. The resultingHW and SW instruction latencies are listed in detail at the bottom of Table 6.5. Figure 6.13d shows the energy-efficiency gains of the LNS clusters w.r.t. the FPU cluster[5].

For complex algorithms with many multiplications, divisions and non-linear functions all LNS designs outperform the FPU up to $4.1\times$ in terms of speedup and energy efficiency (Figure 6.13c and d). The *LNS SF* design which supports energy-efficient intrinsic SIN/ COS instructions even exhibits speedups in the order of $2.3\text{-}5.9\times$ for kernels with trigonometric functions. The DCT-II, for example, can take advantage of fast sine and cosine evaluations while the *FP Exact* design has to call expensive software emulations which take 68 cycles. For benchmarks with many additions and subtractions like GEMM, GEMV and ButterW, the *FP Exact* design is 10% more energy-efficient than *LNS SF* with FMA extensions. In case of ButterW, for example, this drop is caused by data-dependencies in the second order sections of the filter, leading to an increased amount of stalls due to the LNS ADD latency. Note that LNS multiplications can be handled very efficiently in the processor cores with single-cycle integer additions. Therefore, the use of FMA instructions for LNS does not improve efficiency the same way as in FP designs as can be seen in

---

5. It should be noted that a second, *shared* FPU design with an overall complexity of 651 kGE has also been evaluated but not included in this comparison as it was much slower (up to 46%) due to many contentions in the FPU interconnect (up to 96% of accesses resulted in stalls).

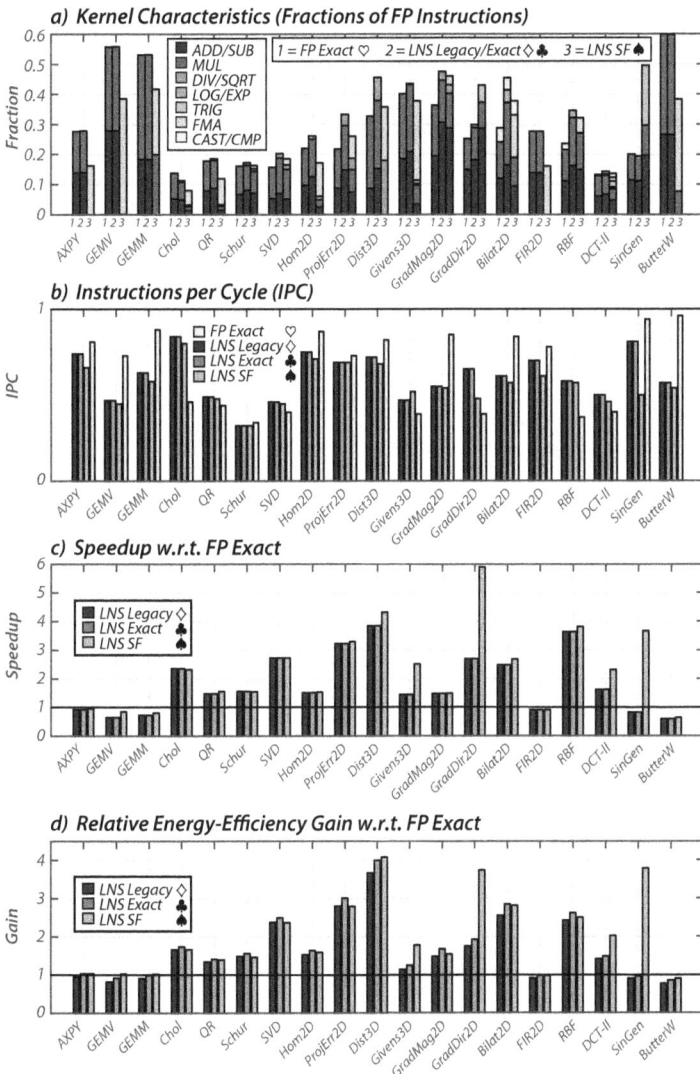

FIGURE 6.13: Application kernel results based on silicon measurements (*FP Exact*, *LNS Legacy*, *LNS SF*) and post-layout simulations (*LNS Exact*): Kernel characteristics in terms of instruction mix (a) and IPC (b). Speedup (c) and energy efficiency (d) gains w.r.t. the FP Exact design.

TABLE 6.7: Description of Function Kernels in the Benchmark.

Kernel	Details
AXPY	BLAS 1 Kernel, [100×1] vectors
GEMV	BLAS 2 Kernel, [10×10] matrices
GEMM	BLAS 3 Kernel, [10×10] matrices
Chol	Cholesky Decomposition, [10×10] matrices
QR	QR Decomposition, [10×10] matrices
Schur	Schur Decomposition, [5×5] matrices
SVD	Singular value decomposition, [5×5] matrices
Hom2D	Evaulation of 2D projective transforms [28, 176]
ProjErr2D	Reprojection error of 2D projective transforms [28, 176]
Dist3D	Distance computations in 3D
Givens3D	Calculation of 3D givens matrix and rotation of 10 vectors
GradMag2D	Computation of gradient magnitude in 2D [175]
GradDir2D	Computation of gradient direction (angle) in 2D [175]
Bilat2D	Evaluation of a bilateral filter in 2D
FIR2D	Evaluation of separable 5×5 Blur Filter in 2D
RBF	Evaluation of a 2D regression function with 25 Gaussian kernels
DCT-II	Evaluation of 1D 32-point DCTs
SinGen	1 kHz sine generator (audio)
ButterW	6th order Butterworth lowpass, 3 SOS (audio)

Figure 6.13c, e.g., for the AXPY, GEMV and ButterW kernels. The utilization of the shared LNU on our benchmarks was 0.42 on average with a maximum of 0.65, leading to an average of 3% stalls due to access contentions (14% in the worst case). For applications where half-precision is acceptable, speedup gains can be increased by 2-2.4× w.r.t. the 32 bit *LNS SF* design, since twice as many operations can be executed per cycle and less stalls occur due to the shorter latencies.

6.9   Summary and Conclusions

The most important findings of this chapter are the following:

- We presented a framework to generate a family of accurate and approximate 1st and 2nd order LNUs capable of executing ADD, SUB, I2F, F2I, EXP, and LOG instructions. The generated LNUs are more compact than most related state-of-the-art designs while providing significantly more functionality.

- While the area cost of a single LNU is difficult to amortize in a single-core processing system, we show that a *shared* LNU can be competitive in size with a traditional private FPU design. Due to the fact that MUL, DIV, SQRT instructions can be efficiently computed in the integer ALUs of the cores, it is sufficient to share one LNU among a cluster of four cores. Despite the fact that additions are more complex and energy-consuming in LNS, we show that the shared LNS designs can compute typical non-linear function kernels up to 4.1× more energy-efficiently than an equivalent chip with four private FPUs.

- We show that the similarity of LNUs with SFUs typically present in today's GPUs allows to conveniently add support for useful transcendental functions (2^x, $\log_2(x)$, $\sin(x)$, $\cos(x)$, atan2(y, x)). Since most of the interpolation resources can be reused, this incurs only a small overhead in the order of 17% for exact 32 bit designs. Function kernels taking advantage of these trigonometric SFEs showed significant speedups in the order of 2.3-5.9× in our evaluations.

- We demonstrate that by relaxing the precision requirements from exact to 16 ULP, significant area savings of 40% can be achieved, and speedups can be further increased up to 5.54× in the best case due to lower LNU latency. Using examples from the image and audio processing domains, we have analyzed the incurred quality losses of the approximate designs in terms of PSNR and THD+N and demonstrated that the errors are imperceptible in the evaluated precision range. We also show that relaxed-precision approximations can be more efficient than simple word width reductions.

- The idea to use LNS as an efficient replacement for FP is not new, and in the light of our findings, the question arises why it has not been deployed on a broader scale so far. There are several reasons for this. First of all, conventional FP is simpler to understand and implement, and it is established in today's compilers and compute platforms, thus discouraging widespread adoption of LNS. Second, standardization is difficult, since LNUs have more degrees of freedom that influence their numerical behaviour (for example all the LUT entries). Consider that – despite standardization of conventional FP arithmetic – there are still cases where different general purpose platforms implementing FP do not exhibit the same numerical behaviour [226, 227]. Therefore, standardization represents an even bigger challenge for LNS that needs to be overcome for widespread adoption.

Given the current trend to incorporate more and more challenging signal processing algorithms into embedded and mobile systems, support for energy-efficient HDR arithmetic becomes ever more important – not only as enabler but also for convenience and rapid application development. Especially for applications involving frequent evaluation of complex non-linear functions, we think that, despite its standardization issues discussed before, LNS arithmetic is a strong contender for HDR arithmetic.

Chapter 7

Stereo-to-Multiview Prototype

7.1 Introduction

Warping-based MVS, as well as alternative approaches, are computationally intensive – yet they should run efficiently in real-time and should be portable to end-user devices to develop their full potential. To this end, we design and implement a complete MVS system that is able to process full-HD content in real-time, leveraging insights gained on sparse features, linear solvers and IDW rendering techniques from Chapters 2, 4, 5 and prior work [33]. In particular, we make the following contributions in this chapter:

- We design and implement the first complete hardware system for IDW-based, automatic MVS. While some subcomponents have been addressed in more detail previously[1] the system integration aspect is new and represents one of the main contributions.

- The IDW processing pipeline for MVS [14] is revisited and the algorithms involved in the substeps are selected such that an efficient HW implementation is possible.

---

1. Point correspondences from S3D content in Chapter 2, saliency estimation in [130], solvers in Chapter 4 and [150], and IDW rendering in [36, 51].

- Based on linear solver architectures discussed in Chapter 4, we design an improved solver which contains a programmable matrix assembly stage and makes use of custom, fused FP arithmetic to significantly reduce the overall pipeline latency and improve on the maximum operating frequency.

- We assess the achievable MVS quality, and provide illustrations of success- and failure cases.

- Using the prototype design, we estimate the implementation cost and energy-efficiency of a completely integrated solution and show that our proposal is viable as an accelerator in a mobile SoC. Also, we provide a comparison with related components from DIBR-based systems.

The chapter is structured as follows: Section 7.2 gives an overview of previous and related work. Next, the processing flow and the choice of the algorithmic components are explained in Section 7.3 which is followed by a description of the corresponding hardware architecture in Section 7.4. The results are presented in Section 7.5, together with a resource estimation of a completely integrated solution.

7.2   Background and Related Work

In this section, we discuss related work on view synthesis as well as related hardware architectures and real-time systems. MVS using IDW relies on a couple of techniques which possess related work of their own. Eventually, all of them have to be considered in order to build a real-time hardware system.

7.2.1   Multiview Synthesis Methods

The IDW method used in this work relies heavily on prior work on warping approaches for video content adaption such as aspect-ratio retargeting [30], S3D retargeting [31] and non-linear disparity mapping [32]. In these applications, each video frame is warped in a content-aware manner in order to fulfill certain constraints. Image features (such as edges) and visual importance maps (saliency) are

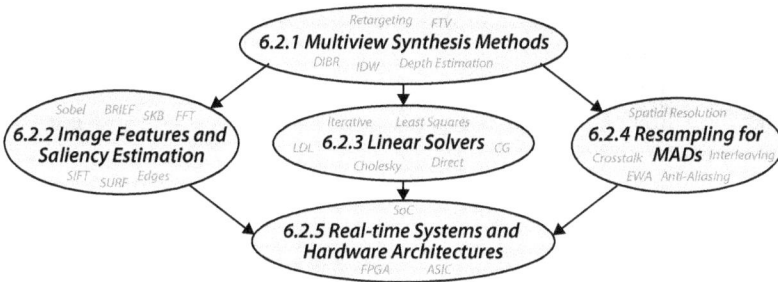

FIGURE 7.1: MVS using IDW relies on a variety of vision and signal processing methods required for a running hardware system.

used in order to determine which parts of the image are important and should be preserved. Unavoidable distortions are moved to visually unimportant regions. This IDW framework has been extended for automatic MVS by [14]. The artifacts caused by geometrical incorrectness are minor and visually hardly noticeable. In fact, exhaustive and formal subjective experiments performed by MPEG [14,228] revealed that IDW as proposed here performs at least as well as DIBR methods, even if those used pre-computed and hand-tuned depth maps. Based on these results, MPEG adopted warp coding for the 3D extension of the new HEVC standard [229].

Related methods [3,7,9,13] for MVS and free-viewpoint television (FTV) rely on DIBR in order to generate new viewpoints. These methods use dense disparity maps to re-project pixels into new virtual views using the relations from epipolar geometry [28]. Although physically correct, DIBR still has two major limitations that remain fundamentally unresolved. First, DIBR-based rendering always introduces disocclusions which have to be filled using inpainting techniques. Second, a dense disparity map of the scene is required which is difficult to obtain in practice, and in real-time implementations often noisy and incomplete.

7.2.2 Image Features and Saliency Estimation

A crucial ingredient for multiview synthesis using IDW is establishing robust sparse point correspondences between the two input images. Numerous algorithms and variations thereof have been devised over the past decade (see Chapter 2 and Chapter 3 for a more in-depth review). Especially so termed *binary* descriptors such as BRIEF [57] and SKB [60] are attractive for hardware implementations since they are inexpensive to compute, require less storage, and can be matched very efficiently. Therefore, our prototype system employs the HW architecture for SKB features developed previously in Chapter 2. Further, the IDW pipeline relies on so-called *saliency maps* that indicate visually important regions in the images. Algorithms for the extraction of saliency information are usually based on psycho-visual attention models [34], and can greatly vary in computational complexity. E.g., a recent, high-quality saliency for S3D was proposed by [230]. Their model not only considers low-level cues such as luminance and chrominance, but also high-level cues. It involves many substeps such as optical-flow calculation, segmentation, and scene-type classification and therefore is less suited for a real-time HW implementation. Here, we use an efficient implementation [130] of the FFT-based saliency proposed by [35] which offers a balanced tradeoff between computational complexity and quality.

7.2.3   Linear Solvers

At the core of most IDW algorithms lies a quadratic energy minimization problem (Section 7.3). Such problems can be tackled by solving a linear system with typically banded sparsity structure. Since this system is positive definite and symmetric, CG- and CHD-based solvers are usually the algorithms of choice [135–137]. The design of HW architectures for these iterative and direct solver types and related work thereof are discussed in detail in Chapter 4, and the analyzed architectures serve as a basis for CHD solver implemented in this system.

7.2.4   Resampling for Multiview Displays

The MAD-specific interleaving process essentially represents a resampling step onto a possibly non-orthogonal or completely irregular sampling lattice. This matter is discussed in detail in Chapter 5 and [5, 6]. Due to the imperfect optics of the MADs, the views often suffer from crosstalk which can be mitigated using appropriate pre-filtering [6]. Further, MADs only have a limited depth budget, and the content to be viewed can be optimized in order to reduce spatial aliasing artifacts [231, 232]. In this work, we use an adapted version of EWA splatting [51] for the warping and interleaving processes. As mentioned in Chapter 5, this is a *forward-mapping* algorithm and its mathematical properties allow to analytically combine interpolation and anti-aliasing filters for the transformation with the display pre-filter.

7.2.5 Real-time Systems and HW Architectures

Most real-time systems for MVS are based on DIBR, and the depth estimation and view synthesis steps are usually treated separately in literature. There are only a few publications where both parts have been combined, such as the work by [233] where one of the first complete real-time systems for MVS has been implemented using a high-end workstation. In [234], an FPGA-based system able to synthesize one synthetic view from 1080p S3D input at 60 fps is presented.

FPGA-based hardware accelerators for depth estimation have been developed by [235–239]. A high-quality, global stereo matching algorithm, achieving a throughput of 720p@2.6 fps is implemented by [235]. The design by [236] outputs CIF depth maps @42 fps whereas the designs by [237] and [238] reach resolutions of up to XGA@60 fps. The architecture from [239] is based on the algorithm used in the MVS system [233] and shows very promising performance of up to 2×1080p@30 fps (S3D). DIBR-based view synthesis engines based on FPGAs and CMOS integrated circuits are developed in [240] and [241–244], respectively. In [242], [244] and [243], single view single view synthesis engines with throughputs of 1080p@32 fps, 1080p@94 fps and 4 k @216 fps, are developed. MVS rendering solutions for 1080p@60 fps and SXGA@29 fps are presented in [240] and [241].

In contrast to mentioned work, we devise a complete, IDW-based hardware system that takes a 1080p S3D input stream and transforms it into interleaved 1080p video with eight synthetic views for display on a MAD. The rendering subsystem of our MVS pipeline has been integrated on an ASIC in 65 nm and published previously in [36]. A more detailed comparison with related work is given in Section 7.5.4.

7.3   Algorithmic Flow

This section is a summary of the algorithmic flow of the implemented MVS scheme, and is based on work presented previously in [14, 150]. Further, we explain the specific selection and parametrization of the methods involved with respect to HW efficiency. As illustrated in Figure 7.2, the pipeline consists of the following five stages:

1. The input to the IDW processing pipeline is the S3D footage (left and right images) which is analyzed in order to reveal image features such as point correspondences, edges and saliency information.
2. Those features are then used to formulate an energy minimization problem, the solution of which results in two warps – one for each input image. These warps describe the (nonlinear) transformation of the input images to a viewing position centered between the two original views.
3. New synthetic views are then generated by first interpolating the two warps to the desired view positions.
4. Then the interpolated warps are used to non-linearly transform the S3D input images.
5. Finally, the views are interleaved in such a way that they can be displayed on the MAD.

The individual steps are explained in more detail below.

7.3.1   Sampling Lattices, Domains and Warps

In the following, we use the terms *sampling lattice* Λ [5] and *domain* $\mathcal{D}$ of the image/warp. The former is a generalization of sampling grids, and may also describe regular, non-orthogonal arrangements. The latter describes the physical dimensions that a certain image or warp spans (measured in pixels). This distinction allows to define coordinate quantities (e.g., warps) or feature maps (e.g., saliency) over a certain physical dimension, but with a different amount of sampling positions (e.g., for subsampled quantities). Non-orthogonal sampling arrangements are important to describe the interleaving patterns of MADs. A sampling lattice is defined as all linear combinations of a set

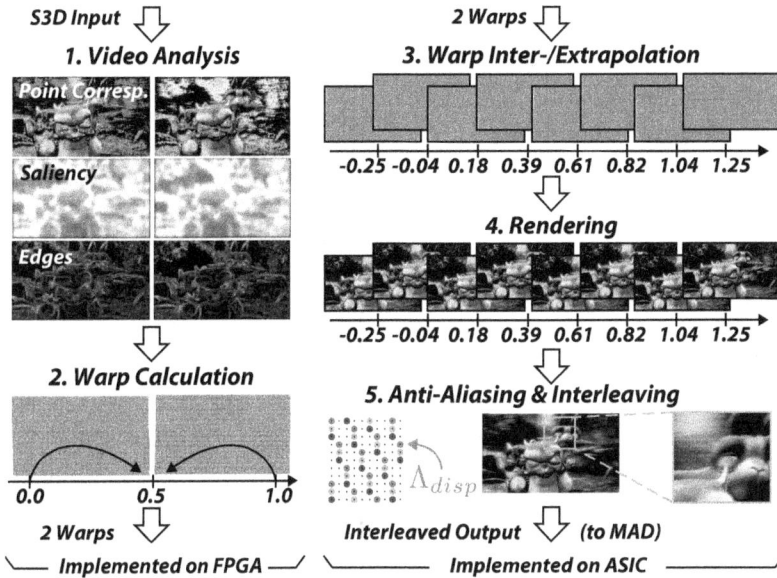

FIGURE 7.2: IDW pipeline for MVS (refer to the text for more details). As indicated, the first part is implemented on an FPGA, and the second part on an ASIC. Images copyright © 2008, BLENDER FOUNDATION, www.bigbuckbunny.org.

of (not necessarily orthogonal) basis vectors $\Lambda = \{\mathbf{u} = n_1\mathbf{v}_1 + n_2\mathbf{v}_2 + \ldots + n_N\mathbf{v}_N | n_i \in \mathbb{Z} = \mathbf{Vn}\}$. To emphasize the distinction between orthogonal and non-orthogonal sampling arrangements, we refer to the former as a *grid*, and to the latter as a *lattice*. Both are defined by a sampling matrix, but in case of a grid we have the restriction that $\mathbf{V}$ is diagonal. A domain of an image or warp is denoted as $\mathcal{D} = \{w \times h\} \subset \mathbb{R}^2$ where w and h are the dimensions in pixels.

An image warp can be described using a non-linear, two-dimensional *forward* mapping $\mathbf{w}(\mathbf{u}) : \mathbf{u} \in \mathcal{D}_{in} \to \mathbf{w}(\mathbf{u}) \in \mathcal{D}_{out}$, where $\mathbf{u}$ is the coordinate in $\mathcal{D}_{in}$. Since it is infeasible to express the image warps as analytical expressions in general, they are stored as discretized functions defined over a two-dimensional grid Λ_{wrp} and with $\mathcal{D}_{in}$. Linearized indices, e.g., k, are used to enumerate the warp vertices $\mathbf{w}_k$ sampled at discrete sampling points $\mathbf{u}_k$ on Λ_{wrp}. These may coincide with the pixel positions in the input image in the case of pixel dense warps. However, this is usually not the case since the warps are often subsampled by around $10\times$ for efficiency reasons (Section 7.3.3). Superscript letters indicate to which view and coordinate-dimension a particular quantity belongs. For example, ^{l} and ^{r} denote the left and right input views, and ^{x} and ^{y} the first and second dimension of a two-dimensional quantity. The two S3D input images will be denoted as $\mathbf{i}^l$, $\mathbf{i}^r$. They are defined over $\mathcal{D}_{in} = \{w_{in} \times h_{in}\}$ and the sampling grid Λ_{in} is defined as $\mathbf{V}_{in} = \text{diag}(1, 1)$.

7.3.2   Video Analysis

Sparse Point Correspondences and Disparities

In MVS, the disparities are the most important features since they reveal the 3D geometry of the observed scene. Yet these have to be robust in order to get good results. As opposed to DIBR-based methods, the IDW approach works on sparse disparities. In this work we adopted features based on SKB [60], since they work well in the setting of almost rectified stereo video. As shown in the evaluations in Chapter 2, SKB features a low outlier rate in S3D settings and, therefore, it is possible to use the features without additional RANSAC filtering, which would be costly in hardware [64]. Occasional outliers can be tolerated since the warp calculation process enforces spatial and temporal smoothness. Furthermore, since SKB is a binary descriptor, it can be calculated and matched very efficiently in hardware. Descriptors containing fixed-point or even FP entries such as SURF [56] or SIFT [55] are much more costly in this respect. Feature points are given in the form of two lists $\mathbf{p}_i^l \in \mathcal{D}_{in}$ and $\mathbf{p}_i^r \in \mathcal{D}_{in}$ with $i \in \{0, 1, ..., n_{pts} - 1\}$. In addition, there is confidence value c_i associated with each pair $\mathbf{p}_i^l$, $\mathbf{p}_i^r$, and we have that $0 \le c_i \le 1.0$.

Saliency Estimation and Edge Maps

A saliency map identifies the visually important regions in the image, and is used to guide the warp calculation such that deformations are hidden in unimportant regions (e.g., homogeneous parts such as blue sky). Extracting visual saliency is difficult since it is a subjective measure that depends to some extent on video content, viewer, and application [34]. Here we use the quaternion Fourier transform

(QFT) algorithm from [35] that exhibits a good tradeoff between computational complexity and quality. The algorithm leverages the *phase-spectrum* of a video sequence, which carries information on where discernible objects are located in an image. Note that the QFT can be efficiently calculated using two separate 2D FFTs [130]. In the following, the saliency maps extracted from $\mathbf{i}^l$ and $\mathbf{i}^r$ are denoted as $\mathbf{s}^l, \mathbf{s}^r$ with values $\in [0.0, ..., 1.0]$, and are defined on $\mathcal{D}_{in}, \Lambda_{wrp}$.

Salient lines should also be preserved in order to avoid bending them in the warped images. This can be done using several methods, e.g., by extracting straight lines using the Hough transform. However, since this feature is not as important as the saliency and the point correspondences, simple gradient-magnitude maps without Canny edge post-processing are used. Such gradient-magnitude maps can be extracted efficiently using 3×3 Sobel filters. The edge maps will be denoted as $\{\mathbf{e}^l, \mathbf{e}^r\} \in [0.0, ..., 1.0]$ and are defined on $\mathcal{D}_{in}, \Lambda_{wrp}$.

7.3.3 Warp Generation

The warps are calculated by solving a quadratic energy minimization problem of the form

$$\min_{\mathbf{f}} \left(E\left(\mathbf{f}\right) \right) = \min_{\mathbf{f}} \left(E_{data}\left(\mathbf{f}\right) + E_{smooth}\left(\mathbf{f}\right) \right) \qquad (7.1)$$

where the data term E_{data} enforces function values at certain coordinate positions, and the smoothness term E_{smooth} is a regularizer that propagates these known values to adjacent sampling positions. The vector $\mathbf{f}$ holds the samples of the unknown warp vertices on Λ_{wrp}. The data term itself is composed of the energy term E_{pt} containing the desired vertex positions in the target image, and the energy term E_t which enforces temporal consistency:

$$E_{data} = \lambda_{pt} E_{pt} + \lambda_t E_t. \qquad (7.2)$$

λ_{pt} and λ_t are weighting parameters to set the relative importance of a particular constraint. Analogously, the smoothness term comprises the two constraints E_{sal} and E_{edge} which determine the local smoothing strength:

$$E_{smooth} = \lambda_{sal} E_{sal} + \lambda_{edge} E_{edge}. \qquad (7.3)$$

Again, λ_{sal} and λ_{edge} are relative importance weights. Recall that we solve for two warps, each of which maps either the left or the right image to a virtual view at position 0.5 on the normalized baseline. This asks for setting up four quadratic energy functionals E^{lx}, E^{ly}, E^{rx} and E^{ry} (two coordinate dimensions for each input view) which can all be minimized independently. In the following we will only describe the constraints assembly for the x-problem of the left warp E^{lx} - the formulations for all other problems follow analogously.

Data Constraints

The term E_{pt} contains the desired positions of the warp vertices in the warped image. As illustrated in Figure 7.2, the target view for both warps is centered between both input views. The sparse point correspondences are used to formulate these constraints by setting

$$ E_{pt}^{lx} = \sum_{i=0}^{n_{pts}-1} \left(s_k^l \cdot c_i^l \right)^2 \cdot \left(f_k^{lx} - 0.5 \left(p_i^{lx} + p_i^{rx} \right) \right)^2, \qquad (7.4) $$

where s_k^l is the saliency value at coordinate position $\mathbf{p}_i^l$, f_k^{lx} is the warp vertex at position $\mathbf{p}_i^l$ and c_i is the associated confidence value. Note that these constraints are only defined for the warp vertices corresponding to $\mathbf{p}_i^l$. There may be many warp vertices without any point constraints. In the case where the warps are subsampled w.r.t. the image resolution, the point constraints have to be distributed to the surrounding four vertices using bilinear weighting [150]. The temporal constraint biases the solution towards the warp vertices $\mathbf{g}_k$ of the previous frame as

$$ E_t^{lx} = \sum_{k \in \Lambda_{wrp}} \left(f_k^{lx} - g_k^{lx} \right)^2. \qquad (7.5) $$

Smoothness Constraints

The saliency constraints

$$ E_{sal}^{lx} = \sum_{k \in \Lambda_{wrp}} s_k^{l\,2} \cdot \left(\left(f_{k+1}^{lx} - f_k^{lx} \right)^2 + \left(f_{k+h}^{lx} - f_k^{lx} - d_x \right)^2 \right), \qquad (7.6) $$

penalize deformations of the regular warp grid (assuming a column-major enumeration k of the warp vertices). d_x denotes the sample spacing, and is related to the warp grid sampling matrix $\mathbf{V}_{wrp} = \mathrm{diag}\,(d_x, d_y)$. The edge constraints penalize deformations orthogonal to the edge directions:

$$E_{edge}^{lx} = \sum_{k \in \Lambda_{wrp}} {e_k^l}^2 \cdot \left(f_{k+1}^{lx} - f_k^{lx} \right)^2. \qquad (7.7)$$

Sparse Linear System

Since the $E(\mathbf{f})$ is quadratic in the elements of $\mathbf{f}$, the solution of (7.1) is an LS solution. As shown in [150], the terms of $E(\mathbf{f})$ can be arranged such that $E\left(\mathbf{f}\right) = ||\mathbf{Af} - \mathbf{b}||^2$, where matrix $\mathbf{A}$ and vector $\mathbf{b}$ both encode the constraints explained before. The global minimum is then achieved if $\nabla E\left(\mathbf{f}\right) = 2\mathbf{A}^T\left(\mathbf{Af} - \mathbf{b}\right) = 0$, i.e. the solution can be found by solving the *normal equations* $\left(\mathbf{A}^T\mathbf{A}\right)\mathbf{f} = \mathbf{A}^T\mathbf{b}$, where $\left(\mathbf{A}^T\mathbf{A}\right)$ is symmetric, quadratic and positive definite. As shown in [150], it is possible to deduce analytical expressions for $\mathbf{A}^T\mathbf{A}$ and $\mathbf{A}^T\mathbf{b}$, which makes the assembly of these equations more efficient.

Since the constraints are defined on small, local neighborhoods on Λ_{wrp}, the matrix $\left(\mathbf{A}^T\mathbf{A}\right)$ is very sparse and only contains one main- and nine off-diagonals. Further, its dimension $n = w_{wrp} \times h_{wrp}$ is in the order of tens of thousands to millions – depending on the resolution of Λ_{wrp}. Current video content is predominantly in 1920×1080 format, resulting in 4 problems (E^{lx},E^{ly},E^{rx} and E^{ry}) with nearly two million variables for each video frame. Solving such large systems at frame rates of up to 30 fps is computationally very demanding and infeasible for real-time applications. As will be shown later in Section 7.5.1, such pixel dense solutions are not necessary as even 10× subsampled grids result in sufficient synthesis quality on many HD S3D sequences. This results in realistic grid sizes of about 180×100, which corresponds to 18 k variables.

Solution Methods

The choice of the solver is dependent on several factors such as the mathematical properties and structure of $\left(\mathbf{A}^T\mathbf{A}\right)$, convergence and numerical behaviour, and the complexity of arithmetic operations and memory accesses. The two most widely used methods for positive-definite matrices are direct CHD- and iterative CG solvers [135–137]. The former employ a matrix decomposition to compute an exact result, whereas the latter successively refine an approximate solution.

As elaborated in more detail in Chapter 4, direct solvers can still be a preferable option over iterative ones even for linear systems with as many as 18 k variables. The reason for this is that iterative solvers require a large memory bandwidth, and may have variable runtime for differently conditioned matrices. E.g., for IDW problems with $n =$ 18 k variables, a single-precision CG solver would have to traverse nine n-dimensional vectors per iteration, and well defined problems require in the order of 200 iterations. This would translate into a memory bandwidth in the order of $30\,\text{fps} \cdot 4 \cdot 18\,\text{k} \cdot 200 \cdot 9 \cdot 4\,\text{Byte} \approx 15.6\,\text{GB/s}$ and memory requirements of around $18\,\text{k} \cdot 9 \cdot 4\,\text{Byte} \approx 5.2\,\text{Mbit}$. Such a large memory bandwidth is difficult to realize with external memory on most embedded platforms, which means that the whole solver would have to be implemented with on-chip resources only, which is very costly in this case. Instead, a direct CHD-based solver only requires two passes: in the first pass the matrix is decomposed into a lower triangular matrix $\mathbf{L}$, a *forward substitution* $\mathbf{h} = \mathbf{L}^{-1}\mathbf{b}$ is performed, and $\mathbf{L}$ and $\mathbf{h}$ are written to the memory. In a second pass, $\mathbf{L}$ is read in reverse order to perform the *backward-substitution* $\mathbf{f} = \mathbf{L}^{-T}\mathbf{h}$. In this case, $\mathbf{L}$ is a banded matrix with 102 nonzero (off-)diagonals, so the external memory bandwidth would be $30\,\text{fps} \cdot 4 \cdot 2 \cdot 18\,\text{k} \cdot (102 + 1) \cdot 4\,\text{Byte} \approx 1.7\,\text{GB/s}$, and the memory requirements for local buffering amounts to only 325.1 kbit.

Based on these numbers we decided to use a CHD-based solver in our hardware implementation. Note however, that for larger systems with several hundred-thousand variables, direct solvers are not viable anymore because of limited parallelism and increased storage requirements for the factorized matrix $\mathbf{L}$. Preconditioned iterative methods offer better tradeoffs for such scenarios, due to better parallelization opportunities and less storage requirements (see also Chapter 4).

7.3.4 Warp Interpolation and Rendering

In the rendering step, the two warps $\mathbf{w}^l$ and $\mathbf{w}^r$ are first bilinearly upsampled from Λ_{wrp} to Λ_{in} in order to form pixel dense warps $\tilde{\mathbf{w}}^l$ and $\tilde{\mathbf{w}}^r$. These are then linearly inter- and extrapolated to all desired view positions α on the normalized baseline. As shown in Figure 7.2, eight views are generated in this case[2]: four views from the left image $\mathbf{i}^l$, and the remaining four views from the right image $\mathbf{i}^r$. The resampling is performed using EWA splatting, which is a *forward-mapping* method. Although more complex than traditional bilinear backward mapping [76], EWA splatting has the advantage that no warp inversion is required since these are calculated in forward format in this application. The EWA framework uses Gaussian filter kernels and the Jacobian of the image warp as a local deformation measure in order to calculate the footprint of an input image pixel in the output image. The input pixels thus correspond to Gaussian *splats* in the output image, which are rasterized within a bounding box and accumulated in a frame buffer. Since Gaussians are closed among themselves and under affine transformations, an anti-aliasing filter for the output image sampling grid can be easily incorporated analytically. A short summary is given below – for a complete derivation see [51].

The EWA Filter Kernel

In the following, $\tilde{\mathbf{w}}(\mathbf{u})$ can be any of the two pixel dense image warps, depending on which view has to be rendered. Let $\mathbf{J}_k$ be the Jacobian of the warp at pixel position $\mathbf{u}_k$. The EWA kernel is characterized by the covariance matrix $\boldsymbol{\Sigma}_{\mathbf{k}} = \mathbf{J}_{\mathbf{k}} \mathbf{W}_{\mathbf{i}} \mathbf{J}_{\mathbf{k}}^{\mathbf{T}} + \mathbf{W}_{\mathbf{aa}}$ in the target image domain, where the first term is the transformed interpolation kernel, and the second term is the anti-aliasing kernel. $\mathbf{W}_i = \text{diag}(\sigma_i^2, \sigma_i^2)$ and $\mathbf{W}_{aa} = \text{diag}(\sigma_{aa}^2, \sigma_{aa}^2)$ are diagonal matrices that parameterize the interpolation and anti-aliasing filters. The Gaussian filter weight at the position $\mathbf{v}_j \in \Lambda_{out}$ on the output sampling lattice is given by

$$\rho_{jk} = \frac{|\mathbf{J}_k|}{2\pi\sqrt{|\boldsymbol{\Sigma}_{\mathbf{k}}|}} \exp\left(-\frac{(\mathbf{v}_j - \tilde{\mathbf{w}}(\mathbf{u}_k))\,\boldsymbol{\Sigma}_{\mathbf{k}}^{-1}\,(\mathbf{v_j} - \tilde{\mathbf{w}}(\mathbf{u_k}))}{2} \right), \quad (7.8)$$

---

2. Note that this view interpolation can be used to linearly scale the depth range of the scene at runtime (see also Figure 7.12 in Section 7.5).

and is multiplied with the corresponding pixel value $\mathbf{i}(\mathbf{u}_k)$. Since the individual transformation of the input pixels can lead to normalization problems in the output image, a post-normalization is required. This is achieved by accumulating the weights ρ_{jk} along with the pixel values $\rho_{jk} \cdot \mathbf{i}(\mathbf{u}_k)$, such that the output pixels $\mathbf{o}(\mathbf{v}_j)$ (defined on $\mathcal{D}_{out}$, Λ_{out}) can be calculated by dividing the accumulated values by the corresponding weights:

$$\mathbf{o}(\mathbf{v}_j) = \frac{\sum\limits_{\forall \mathbf{u}_k \in \mathcal{D}_{in}} \rho_{jk} \cdot \mathbf{i}(\mathbf{u}_k)}{\sum\limits_{\forall \mathbf{u}_k \in \mathcal{D}_{in}} \rho_{jk}}. \tag{7.9}$$

Filter Parametrization and Display Anti-Aliasing

Greisen et al. [51] show that for a regular, quadratic sampling grid the filter parametrization $\sigma_i \approx 0.39$ leads to the optimal L2 fit of a Gaussian to the ideal low-pass filter in the frequency domain. It is also shown how σ_{aa}^2 can be chosen adaptively such that anti-aliasing is only performed when needed. In our application, the covariance matrices are diagonally dominant, and therefore we can use the *simplified adaptive* scheme described by Greisen et al..

The resampled views are interleaved according to a special interleaving pattern (like in Figure 7.2), such that they can be displayed simultaneously on a MAD. Proper care must be taken in order to prevent aliasing, as shown by Konrad et al. [5]. The filters are generally non-separable, and in theory a high order is required to approximate the intricate shape of the passband. But Konrad et al. also noted that for natural images, the benefit of such filters is rather small. As a result, simpler separable filters that lead to visually pleasing results could be used as well. In this work we use the closedness of Gaussians in order to incorporate a Gaussian display pre-filter analytically into the EWA kernel. Instead of using $\sigma_{aa} = 0.39$ we adapt this value with the *density* $1/|\mathbf{V}_{disp}|$ of the display sampling lattice for one particular view as $\sigma_{disp}^2 = \sigma_{aa}^2 \cdot |\mathbf{V}_{disp}|$, where σ_{disp}^2 is now used in place of σ_{aa}^2. The sampling lattices of all views and colors of the display we used for experiments have a density of $1/8$, which results in $\sigma_{out}^2 \approx 1.22$.

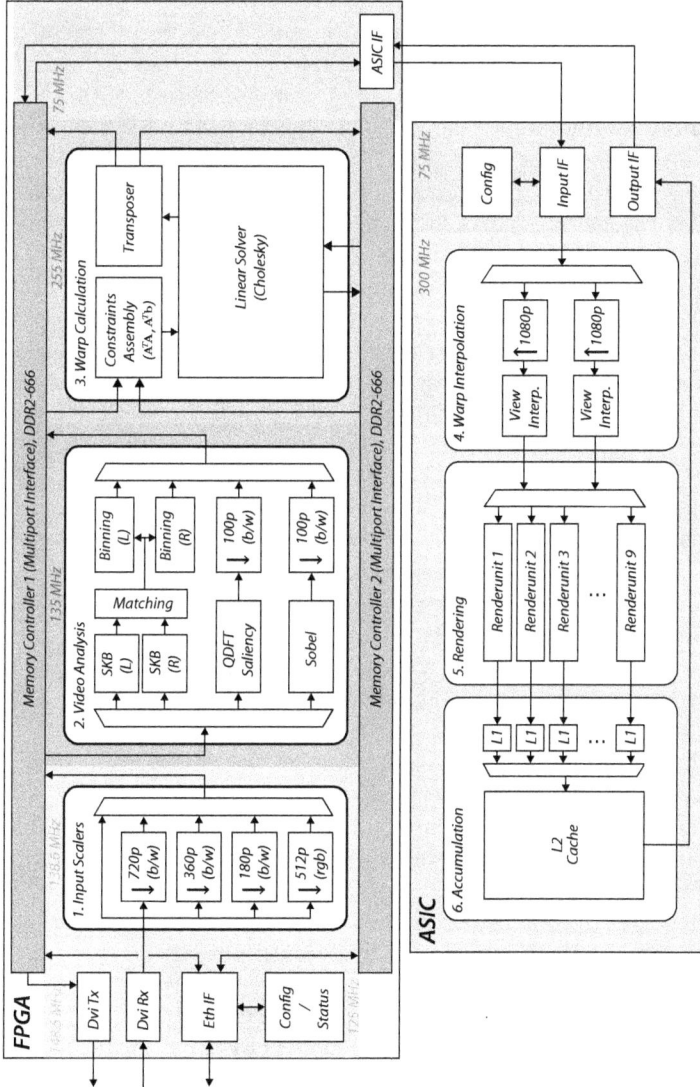

FIGURE 7.3: Top-level blockdiagram of the MVS system. It consists of 6 main parts, where the first 3 reside on the FPGA and the remaining ones on the ASIC: An input scaler stage (1); which preprocesses the input video; a video analysis stage (2), that extracts the features; a warp calculation stage (3), that solves four warps per frame; a warp interpolation stage(4), which performs the warp upsampling and view interpolation; and accumulation (6) stages that resample and interleave the two input images according to the interpolated warps. Each clock-domain has been given a separate colour.

7.4   Hardware Architecture

The complete MVS system shown in Figure 7.3 has been developed in several steps. Initial work concentrated on the design of an efficient MV rendering core in HW which was fabricated as an ASIC in 65 nm. In later project stages, the development was moved to the FPGA prototyping system TERASIC DE4-530. The ASIC was connected via one of the extension ports using a custom extension PCB, and the missing parts have been developed and implemented on the FPGA. The system has been designed to provide enough throughput to process 1080p S3D input video @30 Hz using 180×100 warps. Figure 7.3 provides a high-level block diagram, where the core view synthesis components are numbered from 1 through 6. These components are, in principle, device-independent and can be ported to other FPGA/ASIC technologies. These system processes the input video according to the algorithmic flow presented in Section 7.3:

1. The input scaler block scales the video to the different resolutions that are required later on, and stores them in memory.
2. The video analysis block extracts saliency- and edge maps, and calculates point correspondences using the scaled video frames.
3. The warp calculation block contains a constraints assembly core and a CHD-based linear solver. The former is a programmable unit that builds the LS-problem matrices, and the latter is a fixed function solver.
4. The warp interpolation stage upsamples the two warps to 1080p, and interpolates them to the desired view positions.
5. The rendering block prepares a filter kernel for each input pixel/ warp vertex pair and rasterizes it on the sampling lattice Λ_{disp} of the assigned view.
6. The rasterized subpixels are then sent to the accumulation stage, which fuses them to form the interleaved output image.

The main contributions of this chapter are the complete MVS system shown in Figure 7.3, the programmable matrix assembly stage and the improved linear solver (3). The SKB subsystem in the video analysis block (2) is based on the system explained in Chapter 2, which has been enhanced with feature point sorting stages ('binning' blocks in

Figure 7.3). The saliency implementation in (2) has been taken from [130], and the rendering ASIC comprising (4-6) has been developed previously in [36]. We will concentrate on the new components in the following, while the remaining parts will just be briefly summarized.

7.4.1 Interfaces

The in- and output video streams are both transferred via standard DVI interfaces, referred to as DVI RX and TX. The DVI RX interface is configured to receive S3D video in top/bottom format at 30 Hz with a pixel clock of 138.6 MHz. This allows to conveniently mount the MVS system on any PC or laptop, and play the S3D content using a standard media player. The DVI TX interface is configured to transmit a standard 1080p stream at 60 Hz with a pixel clock of 148.5 MHz. The frame rate of 60 Hz is a requirement of the ALIOSCOPY display we used for testing, since 1080p at 30 Hz is not a supported video standard. As the system is designed for a throughput of 30 fps, the rendered output has to be temporally upsampled and resynchronized. This is achieved by duplicating frames, and a triple buffering scheme is employed to guarantee synchronized frame changes.

Two memory controllers interfacing to a DDR 2-666 DIMM are used - providing a theoretical maximum throughput of 5'333 MB/s each. The controllers are attached to command-based multi-port interfaces employing a simple round robin arbitration policy. The multi-port interface works at $0.25 \cdot 666$ MHz ≈ 166 MHz and with 256 bit wide words such that the throughput is matched. Dual-clock FIFOs are employed in order to facilitate synchronization between the clock domains shown in Figure 7.3. The ASIC rendering core is operating at 300 MHz in order to achieve enough throughput for 30 fps. The I/Os between the FPGA and the ASIC are running at a phase-synchronous clock four times slower (75 MHz) than the core frequency. The ASIC has three 24 bit RGB ports – two at the input and one at the output. The two input images and warps are streamed in through the input ports in an interleaved manner. External memories and status/configuration registers can be accessed from a MATLAB environment via Ethernet in order to facilitate development and debugging.

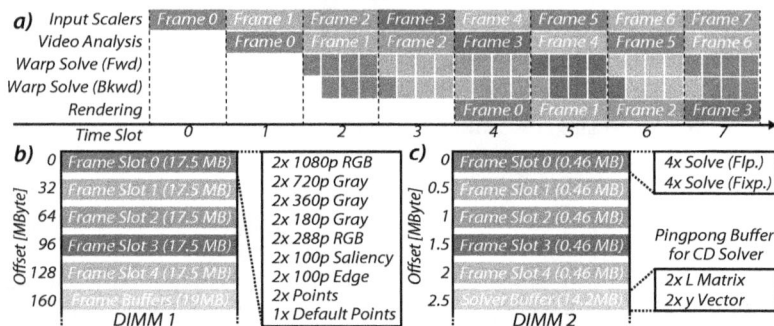

FIGURE 7.4: Processing schedule (a), and memory maps for both DIMMs (b,c). The system works in frame-pipelined manner, and data is exchanged via the external memory. Note, that the solver has a forward- and a backward pass, for which two frame slots have to be allocated as it takes 1.25 frames to solve four problems in a row.

7.4.2   Schedule and Memory Maps

A frame-pipelined architecture was chosen as each HW component has a different requirement in accessing the memory. Some of the blocks process the two images of one S3D frame sequentially, and some in parallel. Further, the solver works on transposed problems, following column-major order as opposed to the row-major order of the image pixel streams. Also, some of the matrix arrays in the solver have to be accessed in reverse order (backwards). In such a setting, it is almost impossible to connect all blocks directly to each other. As shown in Figure 7.4, each component (1-6) works independently on a particular frame, resulting in an overall latency of five frames. The associated data is stored in frame slots and the rendered frames are stored in a triple buffer for temporal upsampling. Temporary decomposition data of the CHD solver ($\mathbf{L}$ and $\mathbf{h}$) is stored in a ping-pong buffer. The memory traffic generated by the individual components is listed in Table 7.1. As the DE4-530 platform provides two external DDR 2-666 memories (DIMM 1 and DIMM 2), the traffic has been distributed to both memory controllers for convenience. Solver related traffic is allocated to DIMM 2, whereas everything else is allocated to DIMM 1. This minimizes memory pattern interference, and allows to use simple controllers and multi-port interfaces with round robin scheduling. As shown in Table 7.1, the theoretical bandwidth utilization is only around 30 % of the maximum available bandwidth[3].

TABLE 7.1: Required memory bandwidth (including alignment overheads).

| [MByte/s] | DIMM1 | | | DIMM2 | | |
	R	W	R+W	R	W	R+W
Scalers	-	497	497	-	-	-
Analysis	99	36	136	-	-	-
Warp Calc.	36	-	36	907	913	1'820
Rendering	398	199	597	6	-	6
DVI TX	398	-	398	-	-	-
Total	932	733	**1'664**	913	913	**1'826**

7.4.3 Stereo Video Analysis

SKB Subsystem

As described in more detail in Chapter 2, the SKB subsystem works on an image pyramid of 720p, 360p and 180p grayscale images. Correspondences are detected in three steps: *interest point detection, descriptor calculation* and *descriptor matching*. The first is basically a filter bank that approximates a Laplacian scale-space using box-filters. Extremal points in this space representing well localized regions in the image are identified using non-maximum suppression. In the descriptor calculation step, small support regions around these extrema are convolved with 16 'semantic kernels', whose responses are binarized by comparing them to zero – thereby producing binary 256 bit descriptors. In the matching stage, the descriptors coming from the left and the right view are matched using a greedy, windowed matching procedure. For each descriptor coming from the left image, the Hamming distances of all correspondence candidates in a small matching window in the right image are calculated. The lowest-distance candidate is accepted if the value lies below the matching threshold. The SKB subsystem is able to extract and match up to 25 k descriptors per frame and the coordinate resolution is 1 pixel in the 720p domain. All data formats are fixed-point and have been chosen such that performance is not compromised.

---

3. Note, that even if bandwidth is in high demand, this is in a completely feasible range for modern mobile SoCs which usually have even faster interfaces such as the LPDDR 3-1600 interface, providing a maximum bandwidth of 12'800 MB/s.

Point Correspondence Binning

The matching block can not guarantee that the correspondences are output in-order, and therefore an additional sorting stage is needed at this point. Further, several correspondences may affect the same warp-quad (the square formed by four warp vertices). We accomplish this by allocating a coordinate bin for each quad, and the correspondences are sorted into these bins on-the-fly. Note that the binning is not equal for the left and the right view, and therefore two such blocks are used. Since the SKB subsystem works in scan-line fashion, the sorting blocks only have to keep a sliding window of around five quad rows. In our evaluations we observed that the number of correspondences per bin does usually not exceed four. Therefore, we align our data structure to DDR memory bursts (256 bit), and hence it is able to hold a maximum of six correspondences per bin (superfluous ones are discarded). This leads to a $180 \cdot 100 \cdot 256$ bit ≈ 0.55 MB data structure for each of the two views. Clearly, there exist sparse data structures more memory-efficient than this (such as quad-trees), but the simplicity and regularity of this approach is very convenient for a HW implementation. Further, since the warp-generation follows column-major order (Section 7.4.4), the burst alignment also simplifies transposition and associated address generation.

Saliency and Edge Extraction

The saliency is calculated by first converting the RGB pixel information into a quaternion representation, followed by a transformation to frequency space using a QFT. In the frequency domain, the phase information is extracted by normalizing each quaternion with its magnitude. An inverse QFT transforms this information back into the image domain. Since the QFT is separable, one QFT can be implemented using two 2D FFTs which in turn can be split into four 1D FFTs, where two are along the rows and two are along the columns of the image. This allows to use an iterative datapath containing one 1D FFTs core that performs the QFT by 4 successive applications. However, this also implies several transpositions, which can be costly in terms of memory bandwidth – especially if the 2D array does not fit on-chip entirely. Therefore, the algorithm has been modified to

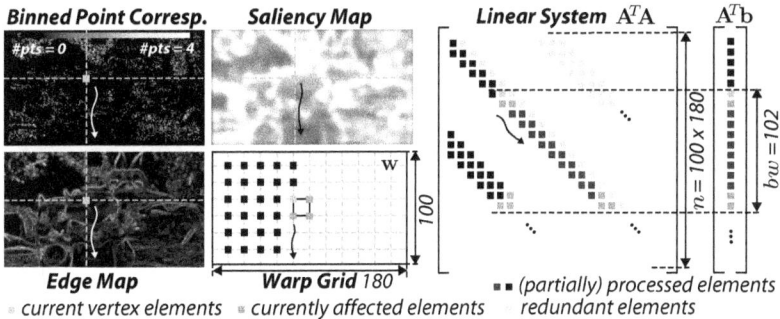

FIGURE 7.5: Constraints assembly: $\mathbf{A}^T\mathbf{A}$, $\mathbf{A}^T\mathbf{b}$ are built along the warp columns. Images copyright © 2008, BLENDER FOUNDATION, www.bigbuckbunny.org.

compute a block-wise saliency, where the image is partitioned into p stripes spanning the full width, but only $1/p$-th of the height. The factor p-can be chosen depending on the on-chip cache size of the target architecture, and has been set to $p = 16$. Border artifacts are mitigated by overlapping the stripes by 25%, and high frequency noise in the saliency map is removed using a 9-tap separable box filter.

The edge maps are calculated using two 3×3 Sobel masks and the vector magnitude approximation from [27], Table 3.8. Clearly, it does not make sense to extract saliency and edge maps from the full resolution 1080p images, since the warp resolution is only 180×100. However, extracting the features from such a low resolution results in poor feature quality – especially in the case where FFTs are involved. Therefore, both the saliency and edge maps are extracted on an intermediate resolution aligned to powers of two (512×288). Then, they are bilinearly downsampled to the warp resolution.

7.4.4 Warp Generation

The assembly of $\mathbf{A}^T\mathbf{A}$ matrices and $\mathbf{A}^T\mathbf{b}$ vectors is sequential in nature and not regular due to the sparse point correspondences. With this in mind, it makes sense to implement it using some sort of programmable architecture. Not only is the development of a program more convenient, but it also simplifies modifications and extensions of the constraints later on. Therefore, we implemented a microcode-programmable mini-processor which is tailored exactly to the needs of the constraints-assembly step. It works on an image feature stream, assembles the constraints on-the-fly, and outputs finished elements of $\mathbf{A}^T\mathbf{A}$ and $\mathbf{A}^T\mathbf{b}$ to the solver.

As discussed in Section 7.3, we use a sparse CHD-based solver, since it offers a good compromise between on-chip memory resources and off-chip bandwidth for problem sizes around $n \approx 18\,\mathrm{k}$. A previous HW implementation of such a solver by Greisen et al. [150] could not reach the required performance of 120 solve/s because of latency overheads incurred by the FP pipeline. Since the CHD has many data dependencies, an FP pipeline deeper than the matrix bandwidth bw (which is 102 for the 180×100 problems at hand) causes many idle cycles where the whole datapath essentially has to wait until all data has propagated to the end of the pipeline. To this end, we make the following two improvements which both aim at reducing the latency below bw and at improving the operating frequency:

- First, the modified $\mathbf{LDL}^T$ Cholesky decomposition is used instead of the standard $\mathbf{LL}^T$ decomposition. This has the advantage that no square-root operator is required [143], which has a high latency (~30 cycles). Further, no divisions are required during backward substitution.
- Second, fused FP arithmetic with partial carry save (PCS) [245] adders is used to implement the large adder tree of the scalar product in the CHD, similarly as this has been done in related implementations for small, dense matrices [246].

Constraints Assembly

The architecture of this block is shown in Figure 7.6: it consists of a feeder unit that loads the image features from off-chip memory, converts them to FP, and feeds them to the microcode-programmable constraints core in *column-major* order. The saliency and edge maps do not occupy a lot of memory, and therefore are loaded at once into the on-chip memory. The correspondence bins are loaded one after another from external memory, and the bilinear weights are determined on-the-fly before being converted to FP.

The constraints core is essentially a FP MAC unit with a 64-entry-register file, a comparator, and a matrix memory. The accumulator is capable of single-cycle accumulation thanks to PCS arithmetic. The

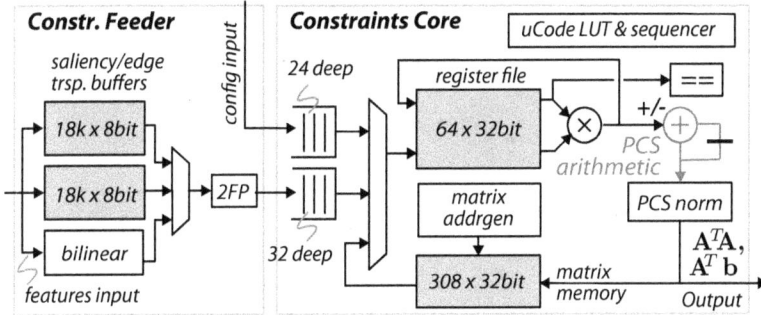

FIGURE 7.6: Architecture of the constraints assembly block. It contains a feeder unit which loads the image features from the external memory, and feeds them to the microcode-programmable constraints core.

microcode instruction set is listed in Figure 7.7 and provides a couple of basic operations, such as MULT, MV, etc. It also supports comparisons and jumps for flow control. Load instructions allow to load data either from the configuration and data FIFOs (LDIN), or from pre-defined relative address offsets in the matrix memory (LDMAT, only the address base changes from vertex to vertex). Move instructions allow to move FP values from register to register, or from a register into the accumulator. Accumulated values can be written back to the matrix memory or output to the solver using store instructions (STMAT, STOUT). For efficient use of the two register file ports, one arithmetic instruction can be parallelized with one load/store instruction.

Once the features have been loaded, the feeder entity loads them into the data FIFO of the core and triggers the execution of the microcode program. This produces one matrix column, for which the core outputs exactly six values: the main diagonal- and the four off-diagonal elements of $\mathbf{A}^T\mathbf{A}$, as well as an element of $\mathbf{A}^T\mathbf{b}$. The execution time of the program has been optimized such that it takes 107 cycles to process one warp vertex in the case where three point correspondences are present (solver processing time for one column is 111 cycles). If no points are present, the program only takes 38 cycles, and for six points it takes 176 cycles. Latency variations are averaged by the input FIFO of the solver. Constants and the parameters λ can be written to the configuration FIFO, which is checked in the initialization phase at the beginning of each frame.

μcode Instruction Set

# variable definition	# no operation,	# load from input FIFOs	# green identifiers are
# variable deletion	# multiplication and	**ldin** config myVar	# predefined addresses /
# (up to 64 vars)	# register move	**ldin** data myVar	# address offsets, e.g. a
var myVar	**nop**		# load from diagA_0 will
del myVar	**mult** a b c	# load from matrix	# load the diagonal element
	mv a b	# memory	# at position (0,0) in the matrix
# static jump,		**ldmat** diagA_0 myVar	
# comparison and	# move into MAC accu		# upon completion of
# conditional jump	**mvmac** a	# store MAC accu into	# a warp vertex, the matrix
jmp label_name:		# matrix memory	# base pointer is incremented
eq myVar 6	# MAC and	**stmat** diagA_0	
jeq label_name:	# signed MAC		# exactly 6 elements are out-
	mac a b	# output MAC accu	# put for each warp vertex (5
# program end	**smac** a b	**stout**	# matrix and 1 vector elements)
end			

FIGURE 7.7: Microcode instruction set of the constraints core.

Cholesky based Linear Solver

The CHD computes a factorization of the form $\mathbf{A}^T\mathbf{A} = \mathbf{L}\mathbf{D}\mathbf{L}^T$, where $\mathbf{D}$ is a diagonal matrix, and the diagonal elements of $\mathbf{L}$ are ones. In order to obtain the solution, we first have to solve $\mathbf{L}\tilde{\mathbf{h}} = \mathbf{A}^T\mathbf{b}$ for $\tilde{\mathbf{h}}$, then we perform the divisions $\mathbf{h} = \mathbf{D}^{-1}\tilde{\mathbf{h}}$ and finally get $\mathbf{f}$ by solving $\mathbf{L}^T\mathbf{f} = \mathbf{h}$. The architecture of the CHD solver is shown in Figure 7.8, and consists of two parts: The first part performs the CHD decomposition and the forward substitution in an interleaved manner, and the second part performs the backward substitution. Although the backward pass is similar to the forward pass, the matrix $\mathbf{L}$ and the vector $\mathbf{h}$ have to be accessed in reversed order. Thus this task can only be executed once the decomposition and forward pass are finished. Using a separate unit is more convenient and allows to perform both passes of two subsequent matrices in parallel.

The decomposition stage contains a wide scalar product, which is used to calculate the inter-row products in column-major order. The parts which use PCS arithmetic have been highlighted. The decomposition is a sequential process since each element $\mathbf{L}_{ij}$ depends on all it's neighbours to the left. However, due to the banded shape of $\mathbf{L}$, the values required to compute another column of $\mathbf{L}$ all lie within a window of size bw^2. These are buffered locally - together with the past bw elements of the $\mathbf{h}$ and $\mathbf{D}$ vectors. Parallelization is easy up to a degree corresponding to bw. Beyond that, the large amount of dependencies on previous results impedes further parallelization.

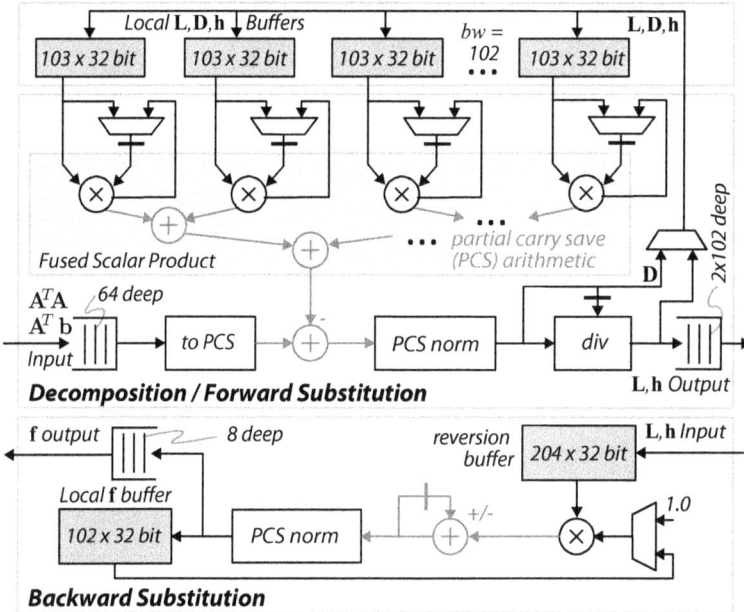

FIGURE 7.8: Architecture of the Cholesky $\mathbf{LDL}^T$ solver. The forward substitution part (top) basically consists of a large scalar product, implemented using a PCS adder tree to minimize datapath latency.

Here, we use a scalar product width of $bw = 102$ in order to meet the throughput requirements. Together with all overheads, the solver requires 111 cycles to process one matrix column, which corresponds to $30\,\mathrm{fps} \cdot 4 \cdot 18\,\mathrm{k} \cdot 111$ cycles $\approx 239.8\,\mathrm{Mcycle}$ per second. Our implementation has some margin and is clocked at $255\,\mathrm{MHz}$.

Numerical Precision and PCS Arithmetic

The FP format employed has been tailored to the precision requirements of this application. Since we calculate coordinate values, the result should be precise to at least ~0.5 pixel so that no artifacts are visible in the rendered images. Using numerical evaluations (Figure 7.9), we decided to use an unsymmetrically biased format (exponent is offset by +8), with 6 exponent and 24 mantissa bits, and an explicitly coded zero bit. One FP word is therefore aligned to 32 bit, and the accumulator length of the PCS adder tree is

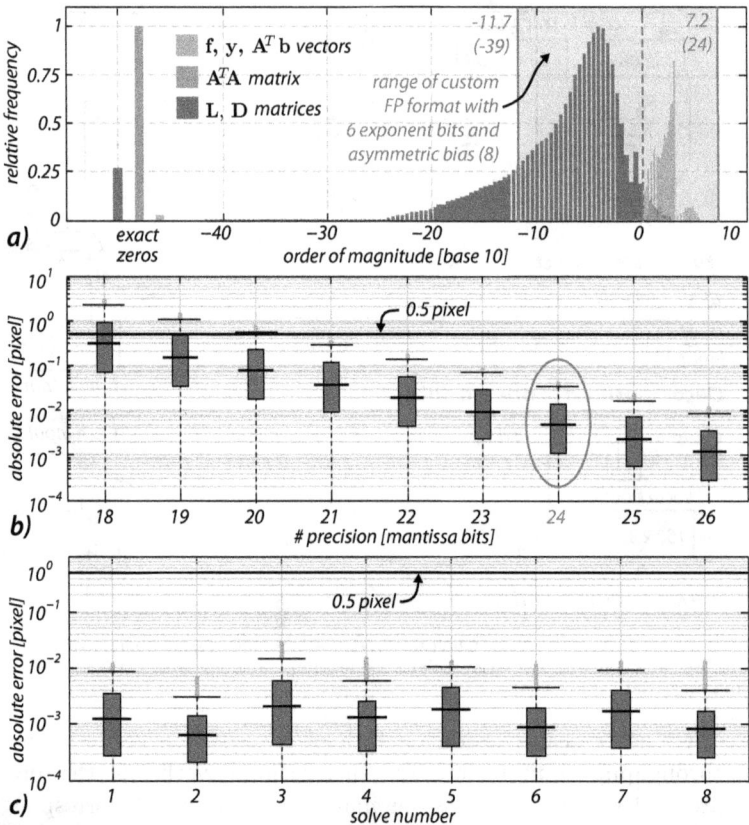

FIGURE 7.9: A custom FP format has been defined by first fixing the exponent range, then the mantissa precision. The histograms in a) show typical magnitude distributions for the quantities involved. The boxplots in b) show the error statistics in dependency of the mantissa precision. Exponent and mantissa have been chosen to have 6 bit and 24 bit, respectively. Error statistics of several different solutions are shown in c).

$24+2^6+1+7=96$ bit (mantissa, exponent range, sign bit and 7 overflow bits). No rounding is performed in any of the FP operators. Test syntheses revealed that 9 bit or less should be used per PCS segment such that 255 MHz can be reached on the target FPGA. With an accumulator length of 96 bit, this corresponds to 11 segments, and therefore PCS normalization requires around 10 cycles.

7.4.5 Interpolation, Rendering and Accumulation

The two input warps $\mathbf{w}^l$, $\mathbf{w}^r$ are first bilinearly upsampled before being interpolated to the desired view positions. After view interpolation, the warp Jacobians are calculated using finite differences. Each warp coordinate is then grouped into packets ('splat jobs') with its Jacobian, its view number and its associated input image pixel. In order to achieve the required throughput the warp interpolation stage has been designed to deliver two splat jobs in each cycle with eight views enabled.

Each render unit contains a filter setup stage, which iteratively prepares the filter kernels, and three rasterization units that evaluate the filter at programmable sampling points on the sampling lattice of the corresponding view. Only the required subpixels of the target image are evaluated, and the rasterizers are designed such as to evaluate one subpixel on the sampling lattice per cycle. Each render unit is able to process one splat job in four cycles, which translates into a throughput of 75 Msplat/s per second. This is sufficient to resample 1080p images at 30 fps as this amounts to 62.2 Msplat/s.

Only the subpixels required are evaluated in the render units and around 2 Gsubpixel/s need to be accumulated per color channel with eight views enabled. This corresponds to ~6.7 subpixel values per color channel and cycle. Fortunately, the large overlap among subsequent splats *of the same view* can be leveraged to reduce this number by placing small, fully-associative subpixel-caches right after the rasterizers (L1 caches in Figure 7.3). These L1 caches reduce the required accumulations by a factor of 5.6 which means that the L2 cache now needs to accumulate around 1.2 subpixels per colour channel and cycle. The L2 cache is the actual framebuffer and is implemented as a sliding window that automatically adjusts its position depending on the incoming addresses. Assumptions on the geometric arrangement (i.e. almost rectified views) of the views allow to store only a small excerpt (25 rows) of the whole output image on chip, and therefore no external memory is required. All internal number formats are fixed-point, and have been tailored such that no quality loss can be perceived in visual comparisons w.r.t. to a double-precision FP implementation of the warp-interpolation and rendering blocks.

7.5 Results

7.5.1 Multiview Synthesis Results

Rendered results are depicted in Figures 7.10 and 7.11, and an example for (real-time) depth-volume adjustment is shown in Figure 7.12. All depicted results have been grabbed from our HW system. During informal subjective tests using full-length S3D movie footage, we found that the system works well on a broad range of synthetic, as well as live action content using the same set of parameters. In general, the spatial artifacts of the IDW method are rather subtle in nature since the image texture is transformed as a whole, and no discontinuities are produced. This is why even rendered images with artifacts can be visually pleasing. Typical artifacts are described in more detail below, and for exhaustive and formal subjective experiments we refer to the study performed by [14,228], which revealed that fully automatic IDW as proposed here performs at least equally well as DIBR methods.

Spatial Artifacts

Typical spatial IDW artifacts are excessively bent image regions, which are often caused by large changes of perspective, large antagonistic disparities in adjacent image regions, multiple disocclusions, or too few point correspondences. Figure 7.10 d-f depict cases where fore- and background have large opposite disparities. In such cases, a change of perspective would usually lead to disocclusion of large image areas, and the IDW method has to handle this by streching/compressing the transition regions.

Enough accurate point correspondences are essential for good performance. Usual failure cases are small or thin foreground objects which do not yield enough correspondences, and therefore are 'smeared' into the background, e.g., Figure 7.10 c1, and Figure 7.11 a-f. Although we do not perform a RANSAC filtering step, there are only very few artifacts due to false correspondences, since the smoothness constraints in the warp generation establish a certain outlier tolerance. Artifacts are sometimes visible in repetitively or homogeneously textured regions, such as in Figure 7.11 g. As shown

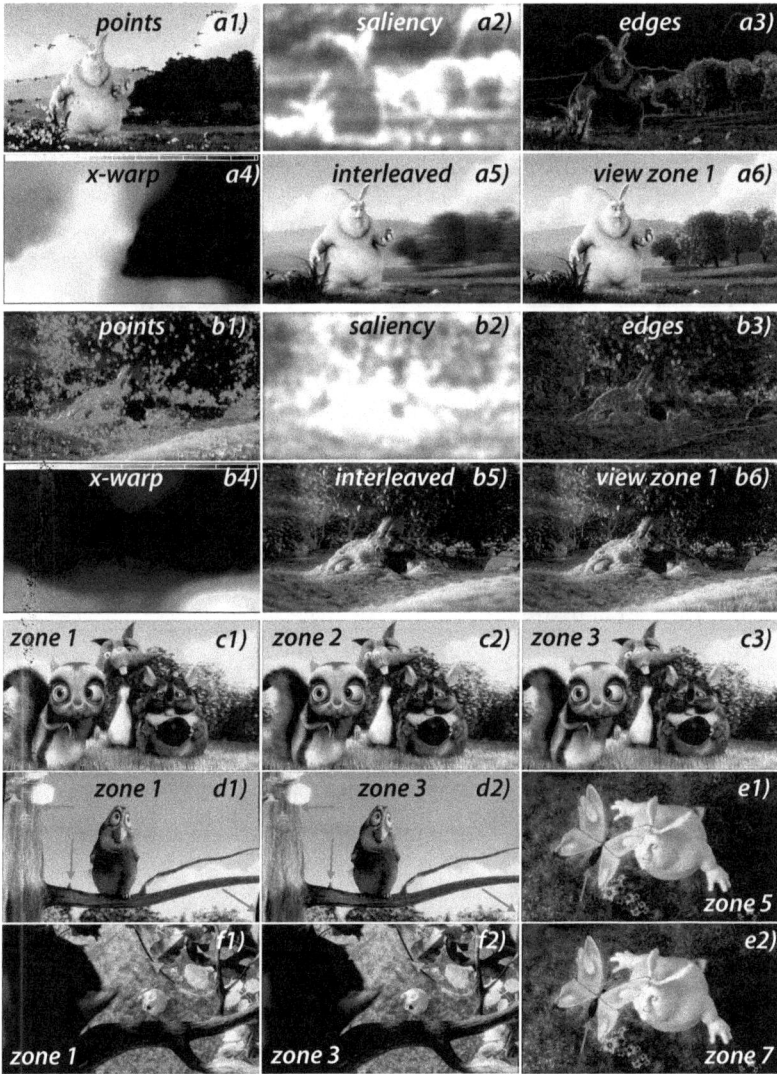

FIGURE 7.10: a-b show two examples with all feature maps of the left view and the rendered output, c1-c4 show three first viewing zones, and d-f show results with annotated warping artifacts. The results are in anaglyph format – use a document viewer to zoom in and view with red/blue glasses. Images copyright © 2008, BLENDER FOUNDATION, www.bigbuckbunny.org.

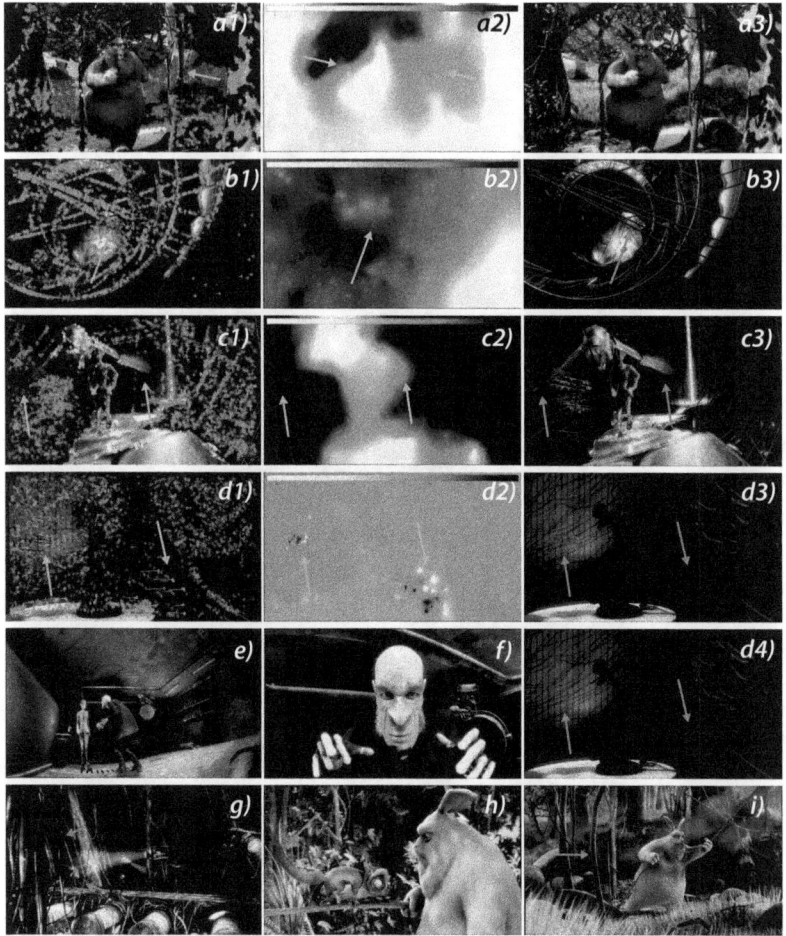

FIGURE 7.11: a-g show results with annotated warping artifacts, and h-i show two additional scenes where the IDW method works well. The results are in anaglyph format – use a document viewer to zoom in and view with red/blue glasses. Images copyright © 2008, BLENDER FOUNDATION, www.bigbuckbunny.organd copyright © 2006, BLENDER FOUNDATION and NIMK, www.elephantsdream.org.

in Figure 7.11 g4, increasing the smoothness weight λ_{sal} can help to reduce such artifacts – however at the price of increased warp stiffness.

The current implementation does not perform any blending of textures coming from left and right views. This has the advantage

FIGURE 7.12: Example for depth-volume adjustment (the percentages indicate the scaling factor of the original baseline). The results are in anaglyph format – use a document viewer to zoom in and view with red/blue glasses. Images copyright © 2008, BLENDER FOUNDATION, www.bigbuckbunny.organd copyright © 2006, BLENDER FOUNDATION and NIMK, www.elephantsdream.org.

that less views need to be rendered, and that these views do not exhibit any ghosting artifacts due to warp misalignment. A series of three adjacent viewing zones is shown in Figure 7.10 c. Note that these views are all slightly different and enable a limited, horizontal head-motion parallax when viewed on a MAD.

Temporal Artifacts

Without any temporal consistency constraint, the warps can exhibit disturbing temporal jittering artifacts (Figure 7.13 a) , since the extracted image features are slightly different from frame to frame. In order to overcome this issue, a temporal regularizer is used which acts as a smoothing filter. However, as can be seen in Figure 7.13 a-b there's obviously a tradeoff to be made between jittering artifacts and over-smoothing which, for large values λ_t, can be visible as temporally lagging depth adjustments. For smaller, more reasonable values of λ_t, the effect of temporal regularization can only be noticed around fast moving image regions or at scene cuts[4].

Warp Resolution and Parameterization

The impact of the warp resolution is illustrated with the example in Figure 7.14 b,c. While small resolutions are computationally preferable, they deliver poor performance since the essential scene geometry is not captured accurately enough. High-resolution warps perform visually better, but they are also much more expensive to compute.

---

4. A scene-cut detector can be employed in order to alleviate this issue.

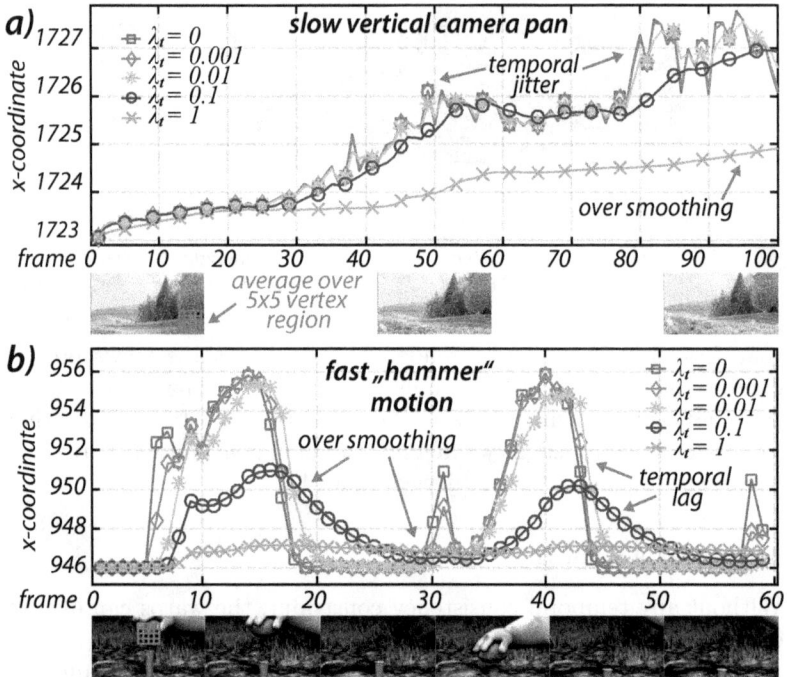

FIGURE 7.13: The effect of temporal regularization is illustrated using a slowly moving scene (a) and a scene with fast motion (b). The plots show how the average x-coordinate of a 5×5 warp patch evolves over time. For small values of λ_t, one can observe temporal jittering artifacts, whereas for large values of λ_t we can observe over-smoothing and a temporal lag in image regions with fast motion (with $\lambda_{pt} = 1$, $\lambda_{sal} = 10$, $\lambda_{edge} = 100$). Images copyright © 2008, BLENDER FOUNDATION, www.bigbuckbunny.org.

A resolution around 180×100 has therefore been found to be a good compromise between visual quality and computational complexity. For higher fidelity, warps in the order of 360×200 could be used in future designs.

There is a range of well-performing parameter combinations and the specific choice is up to the user. We found that setting the (relative) weights to $50 \cdot \lambda_t = 1 \cdot \lambda_{pt}/c_{pt} = 0.1 \cdot \lambda_{sal} = 0.01 \cdot \lambda_{edge} = 1$ provides good results on most content, where the factor $c_{pt} = (180 \cdot 100)/n_{pts}$ accounts for the fact that all constraints except the point correspondences are defined using 2D feature maps.

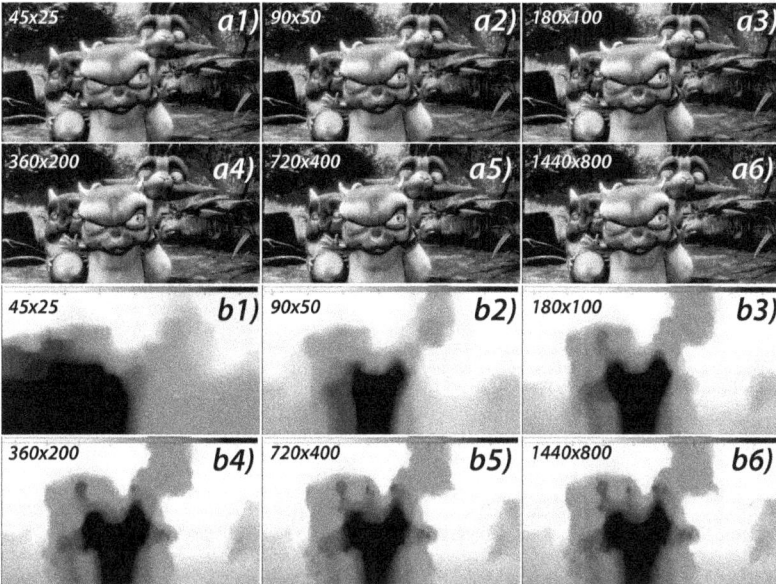

FIGURE 7.14: a1-6 example rendering using different warp resolutions. b1-6 are the corresponding x-warps after bilinear upsampling. Results are in anaglyph format, use a document viewer to zoom in and view with red/blue glasses. Images copyright © 2008, BLENDER FOUNDATION, www.bigbuckbunny.org.

7.5.2 Functional Characteristics and Performance

Table 7.2 provides a summary of the resource utilization of the developed MVS system. The input video is in top/bottom tiled S3D format with 1080p resolution per view and 30 fps. The output is a 1080p 60 Hz video stream, where the individual frames contain eight interleaved images for a MAD (currently for the ALIOSCOPY HD 47" long-view, but can be adapted to similar displays). The system is able to convert S3D video at 30 fps which allows for real-time operation. Further, the warp generation parameters, number of views (up to nine), the position of the views on the normalized baseline and the display parameters (interleaving pattern and filter parametrization) are fully programmable at runtime. This allows, for example, to adjust the displayed depth volume in real time (linear scaling of the camera baseline) as shown in Figure 7.12. Further, our system does not assume completely rectified input content. In fact, a Keystone

TABLE 7.2: Physical characteristics of the multiview system. The FPGA is an ALTERA Stratix IV (EP4SGX530KH40C2), and the ASIC has been fabricated in 65 nm technology. Frequency values represent the current parametrization.

FPGA		Logic			RAM		Freq.
	LUTs	Regs	DSPs	LUTs	9/144K	[MHz]
Scalers	4.1 k	7.1 k	24	2.3 k	10/0	138.6
Analysis	43.3 k	47.3 k	177	2.7 k	434/24	135
1 SKB Core	11 k	11.3 k	32	0	11/0	135
Matching	6.2 k	2.8 k	1	0.3 k	249/4	135
1 Binner	0.5 k	0.9 k	0	0.3 k	31/0	135
Saliency	8.3 k	13.2 k	96	1.1 k	95/0	135
Sobel	0.3 k	0.3 k	0	0	2/0	135
Warp Calc.	51.9 k	67.4 k	420	0.7 k	148/0	255
Assembly	4.8 k	6.6 k	8	0.2 k	38/0	255
Solver	45.8 k	58.9 k	412	0.5 k	104	255
Transposer	1.3 k	1.9 k	0	0	6/0	255
Subtotal	99.3 k	121.8 k	621	5.7 k	592/24	↑
IO/Infra.	22.1 k	35.5 k	0	15.7 k	14/4	N/A
Total	**121.4 k**	**157.3 k**	**621**	**21.4 k**	**606/28**	↑

ASIC		Logic			SRAM		Freq.
	[mm²]	[kGE]	[Mbit]	[mm²]	[kGE]	[MHz]
IO/Infra.	0.28	195	-	-	-	75, 300
Warp Interp.	0.314	218	0.76	1.023	710	300
Rendering	2.317	1'609	-	-	-	300
Renderunit	0.257	179	-	-	-	300
Accumulator	0.385	267	3.6	5.436	3'775	300
Total	**3.296**	**2'289**	**4.36**	**6.459**	**4'485**	↑

distortion [1] of up to ±11 pixel can be tolerated. Note that the same rendering architecture implemented in the ASIC could support quad-full-HD (3840 × 2160) output resolution at the same framerate by increasing the I/O bandwidth of the L2 cache. Only practical I/O limitations (no flip-chip packaging) have prevented us from doing so.

7.5.3   ASIC Complexity and Power Estimates

In order to get an idea of the area and power requirements of a completely integrated system, the major blocks have been ported to 65 nm and 28 nm technology. The results are shown in Table 7.3 and with the exception of the saliency estimation block, all results are

TABLE 7.3: Estimated complexity and power consumption[5] of the major blocks in 65 nm (LVT, TT @1.3 V 25 °C) and 28 nm (LVT, TT @1 V 25 °C).

| ASIC Estimates | Logic | | SRAM | | Power |
	$[mm^2]$	$[kGE]^\diamond$	$[mm^2]$	$[kGE]^\diamond$	[mW]
Scalers*	0.136	94	0.406	282	19
Analysis*	1.099	763	9.414	6'537	343
Warp Calc*	1.841	1'278	1.794	1'245	791
MADMAX ASIC[†]	3.296	2'289	6.459	4'485	752
Total (65 nm)	**6.372**	**4'424**	**18.073**	**12'549**	**1'905**
Total (28 nm)*	**1.444**	**3'966**	**6.929**	**15'069**	**605**

* Gate level synthesis and simulation   † Measurements
$^\diamond$ 1GE = 1.44 μm^2 (65 nm), 1GE = 0.364 μm^2 (28 nm)

based on gate-level synthesis and simulation (SYNOPSYS DC, MENTOR Modelsim and SYNOPSYS Power Analyzer). Due to proprietary FPGA macro dependencies (FFT cores), the values for the saliency block have been estimated using the FPGA and ASIC results of the Cholesky solver (which has similar characteristics in terms of DSP, LUT and register usage). Note that the estimates do not include the infrastructure and interface blocks from the FPGA part.

It is important to keep in mind that the power consumption due to the external memory accesses is not negligible. Using the estimation procedure described in Chapter 4, we calculated that an LPDDR 3 PC6400 memory subsystem would consume around 700 mW supplying a bandwidth of 3.5 GB/s (including memory controller, PHY, LPDDR 3 component and I/O switching power). Adding this to the power estimate for 28 nm results in a total power consumption of ~1.3 W (not accounting for additional I/O and infrastructure circuitry). It is interesting to observe, that at this technology node, the power consumption of the external memory interface is even larger than the power consumption of the hardware accelerator.

---

5. For comparison, the GPU subsystems of APPLE's A7 and NVIDIA's Tegra K1 occupy around 21 mm^2 and 40 mm^2, respectively [22], and the platform power consumption of an A7-based tablet is around 6 W when running a graphics benchmark (offscreen) [247].

7.5.4   Comparison with Related Work

There are relatively few published real-time systems which implement a complete MVS pipeline with video analysis and rendering units. Both parts are usually treated as separate subproblems in the literature. Therefore, we first compare our system against complete systems, and then we compare the two subsystems with related depth image based architectures.

Complete Systems

Complete real-time systems with similar functionality have been presented by Riechert et al. [233] and Liao et al. [234] (see Table 7.4 top). Riechert et al. [233] implemented a system based on a dual processor workstation with two INTEL Xeon 5690 CPU's and two NVIDIA GTX 590 graphics cards. It has almost the same performance as ours and is able to synthesize 8 interleaved views from 1080p S3D content at 24 fps. Clearly, the size of this system, the cost, and power consumption make it unsuitable for integration into consumer devices. Similarly to our system – their rendering engine could also support larger 4 k displays without any decrease in performance, since the views are rendered at higher resolution internally. Liao et al. [234] implement a single-view synthesis pipeline on an ALTERA Stratix III FPGA, which is able to produce depth-adjusted S3D content generated from 1080p S3D input video at 60 fps. The system synthesizes one virtual view and bypasses one of the input views. Since no detailed FPGA results where published, we compare their complexity estimates in 250 nm CMOS technology against our ASIC estimates in Table 7.4. At first sight the implementation by [234] looks to be almost a factor 10 more area-efficient. This has several reasons. First of all, their implementation only renders one virtual view and not up to nine views as in our work and [233]. They also make the assumption that the input video is perfectly rectified, i.e., there is no y-disparity between the left and right image, which simplifies all 2D-problems to 1D-problems in the depth estimation and rendering parts. Yet such rectification is not easily achieved, and therefore systems should be designed to tolerate y-disparities (Keystone [1]). For example, our system is designed to support up to ± 11 pixels of y-disparity.

TABLE 7.4: Comparison with other view synthesis systems, video analysis and rendering cores. The Mbit values represent block ram bits in the case of FPGA designs, and SRAM bits (macros) in the case of ASIC designs.

Full Systems	Type	Input	Maps	Output	Technology	LUT	DSP	Reg	RAM [Mbit]	Bw [GB/s]	f [MHz]	fps
This work	IDW	2×1080p	2×100p	8 in 1080p	ASIC 65 nm	4'424 kGE §			10.5§	3.1‡	≤300	30
Liao [234]	DIBR	2×1080p	2×1080p	2×1080p†	ASIC 250 nm	470 kGE §			1.27§	0	?	60
Riechert [233]	DIBR	2×1080p	2×1080p	8 in 1080p	GPU+CPU	2×Xeon 5690	2×GTX 590		590	?	?	24

Video Analysis	Type	Input	Map Out (Range)	Technology	LUT	DSP	Reg	Ram [Mbit]	Bw [GB/s]	f [MHz]	fps
This work	Warp	2×1080p	2×100p (128)	Stratix IV	142.8k	621	157.3k	6.56	3.1‡°	≤255	30
Werner [239]	Depth	2×1080p	2×1080p (?)	Stratix V	70.7k	625	132.9k	3/6.5□	1.6°	150	30
Akın [238]	Depth	2×XGA	1×XGA (128)	Virtex 5	48k	0	43k	2.3	0.57	190	60
Zhang [237]	Depth	2×XGA	1×XGA (64)	Stratix III	80.6k	257	94.9k	3.77	?	100	60
Perez [235]	Depth	2×720p	1×720p (96)	Virtex 5	23.7k§	120§	17k§	0§	19.2	200§	2.6
Chang [236]	Depth	2×CIF	1×CIF (64)	ASIC 90 nm	562 kGE			0.17	0.06	95	42

View Synthesis	Type	Input	Maps In	Output	Technology	LUT	DSP	Reg	Ram [Mbit]	Bw [GB/s]	f [MHz]	fps
This work	IDW	2×1080p	2×100p	8 in 1080p	ASIC 65 nm	2'289 kGE			4.36	0	300	33.1
Tsung [243]	DIBR	?	?	1×4k	ASIC 40 nm	1'416 kGE			0.16	?	240	216
Horng [242]	DIBR	2×1080p	2×1080p	1×1080p	ASIC 90 nm	269 kGE			0.55	0.96	200	32.4
Chang [244]	DIBR	2×1080p	2×1080p	1×1080p	ASIC 90 nm	143 kGE			0.44	1.27	200	94
Fan [241]	DIBR	?	?	8 in SXGA	ASIC 180 nm	104 kGE ×			0.008	0	71.4	29.1
Chen [240]	DIBR	1×360p°	1×360p°	9 in 1080p	Cyclone III	1.4k△	0	1.3k	0.47	0	148.5	60

° Including bandwidth for S3D bypass □ ROM/RAM ° 640×360 △ 4-input LUTs × Including SRAM
† One bypassed view, one rendered view § Estimates, not accounting for memory controller and other I/O infrastructure.
‡ DVI TX bandwidth due to temporal upsampling has been omitted, since 30 fps could be directly output from the rendering core.

Video Analysis and Depth Estimation Cores

In DIBR frameworks, depth estimators such as the implementations by [235–239] can be viewed as the equivalent part of the video analysis block in an IDW-based system. A quantitative comparison is shown in Table 7.4 in the middle. Note that the design by Perez et al. [235] uses a global, high-quality technique which requires a very large memory bandwidth, and therefore not suited for real-time frame rates. The other designs are all local methods using either a variation of the census-transform [236, 237], a combination of absolute differences and the census-transform [238], or normalized cross-correlation [239]. These local methods are viable for real-time systems due to their higher throughput and lower bandwidth requirements. However, this comes at the cost of more artifacts in the depth maps. Among the listed designs, the one by Werner et al. [239] is the most recent and is best suited for a 1080p MVS system. In fact, it implements a hardware friendly version of the algorithm used in the system by Riechert et al. [233], and is thus tailored to the needs of MVS.

Note that there are several reasons why it is difficult to directly compare the video analysis stage of our IDW system with related depth extraction cores, although both the image warps and depth maps essentially capture the scene geometry. First of all, the depth maps and image warps used in DIBR and IDW lead to inherently different artifacts which are not similarly perceived. Also in terms of hardware complexity, a direct comparison is difficult due to the fact that most DIBR methods are local methods, whereas IDW is a global method requiring an LS-solve. In addition, some of the methods use internal subsampling in order to be more efficient, and convert the calculated maps to full resolution in a final post-processing step. E.g., [239] work with 4× subsampling, and our method works on 180×100 warps which are later upsampled to 1080p in the rendering ASIC.

View Synthesis Cores

A comparison of the rendering ASIC with related DIBR rendering architectures is given in Table 7.4 at the bottom. [242–244] develop single (non-interleaved) view synthesis engines with performances of up to 1080p@32.4 fps, 1080p@94 fps and 4 k @216 fps, respectively.

In contrast, [240] and [241] develop architectures which are able to directly render interleaved views. As we can see, the hardware complexities of these DIBR architectures vary significantly. A main reason for this is that the designs [240,241,244] make the same assumptions as Liao et al. [234] and work with perfectly rectified content. This allows to collapse the 2D warping operations to 1D problems which can be implemented very efficiently with streaming architectures. Additional reasons for the varying complexity are the different image and depth map resolutions.

When comparing our chip with these related DIBR designs we observe that it is has much larger logic and SRAM area than all other designs. A main reason for this is that we use a forward mapping algorithm, and its computational complexity is defined by the input image resolution. In terms of throughput, our design could therefore easily generate interleaved output images up to the point where each sub-view has 2 MPixel resolution (interleaved 4 k would therefore be possible as well). Only practical I/O limitations prevented us from doing so. Also, we do not make the assumption of perfectly rectified content, which would allow us to simplify the rendering architecture significantly as well. The large SRAM area is due to the warp (0.76 Mbit) and frame buffers (3.6 Mbit). The former has been incorporated for convenience during development and debugging, and is not strictly required since the design works in scan-line order. The latter has been added such that no off-chip memory is required.

7.6 Summary and Conclusions

In summary, we make the following findings:

- We described a HW architecture and implementation of a complete, IDW-based MVS system. Our system is able to synthesize high quality, full-HD content with up to eight views in real-time, and supports a Keystone distortion of up to ± 11 pixels. Further, no inpainting steps are required due to the use of IDW. The current test-bed is designed using an FPGA platform combined with a custom ASIC, and our results show that a monolithic integration of the developed hardware IP into a SoC fabricated in 28 nm is completely feasible and would amount to around $8.5\,mm^2$ (including 4 MGE logic and 10.5 Mbit SRAM).

- With an estimated power consumption of ~1.3 W the hardware accelerator enables portable and energy efficient MVS, which are both essential properties when considering a deployment in consumer electronic devices.

- During informal subjective tests using full-length S3D movie footage, we found that the system works well on a broad range of synthetic, as well as live action content using the same set of parameters.

- This system is the first of its kind. The only other published complete system comprising similar functionality in terms of video analysis and MVS has been implemented on a high-end workstation. When comparing subcomponents of our system to related DIBR-based modules, we observe that some of the DIBR components are more compact in terms of logic and SRAM area. One of the main reasons for this is the fact that several related DIBR components assume perfectly rectified S3D input, which allows to reduce several 2D problems to 1D problems which are easier to implement. Our system and its subcomponents have been designed to tolerate small Keystone distortions, and there-fore operate on 2D problems. So in summary, these comparisons should be interpreted with caution, as the algorithms involved are different and the synthesis quality has not been part of this comparison.

There are several aspects with potential for future improvements:

- The system has – in principle – already enough rendering throug-hput to produce 4 k output for newer MAD models. Modifi-cations only have to be made at the output of the rendering subsystem. As proposed in Chapter 5, the rendering part could be optimized further by performing a large part of the display anti-aliasing using a *pre-filter*.

- Since most IDW pipelines are similar, the system could in prin-ciple be extended to support other applications such as video retargeting [30], non-linear disparity mapping [32], retargeting of S3D content [31], or optimum multiview content creation methods [232]. The main part that would need modifications is the assembly of the LS system since these applications mainly differ in the way the constraints are assembled. This underlines the need for a programmable warp assembly stage we have implemented.

- An additional RANSAC step could improve the quality of the point correspondences – yet we found that this is not required on the vast majority of S3D content (only repetitive and homogeneous textures sometimes have issues). Also, the density of the correspondence points could be increased by employing a patchmatch-based point extraction method as used in Chapter 3 for optical flow estimation. Such methods can deliver accurate, quasi-dense NNFs, which have better coverage than interestpoint-detector-based approaches. However, these methods are also computationally more demanding and require more memory bandwidth due to the random search propagations.

- Although all three stages (analysis, warp solver, and rendering) seem to have similar complexity, the computational bottleneck of IDW-based MVS is clearly the linear solver since its complexity does not scale as well as for the other parts when moving to higher resolutions. As discussed in more detail in Chapter 4, there are several alternative approaches that could be explored for higher warp resolutions, such as for example preconditioned, iterative CG solvers.

- As demonstrated in Chapter 3, STEA filters can be used to produce temporally smooth approximations to various regularization problems. It would be interesting to investigate in more detail whether the LS problem at hand can be replaced by a STEA filter, since they have the potential to offer considerable improvements in scalability.

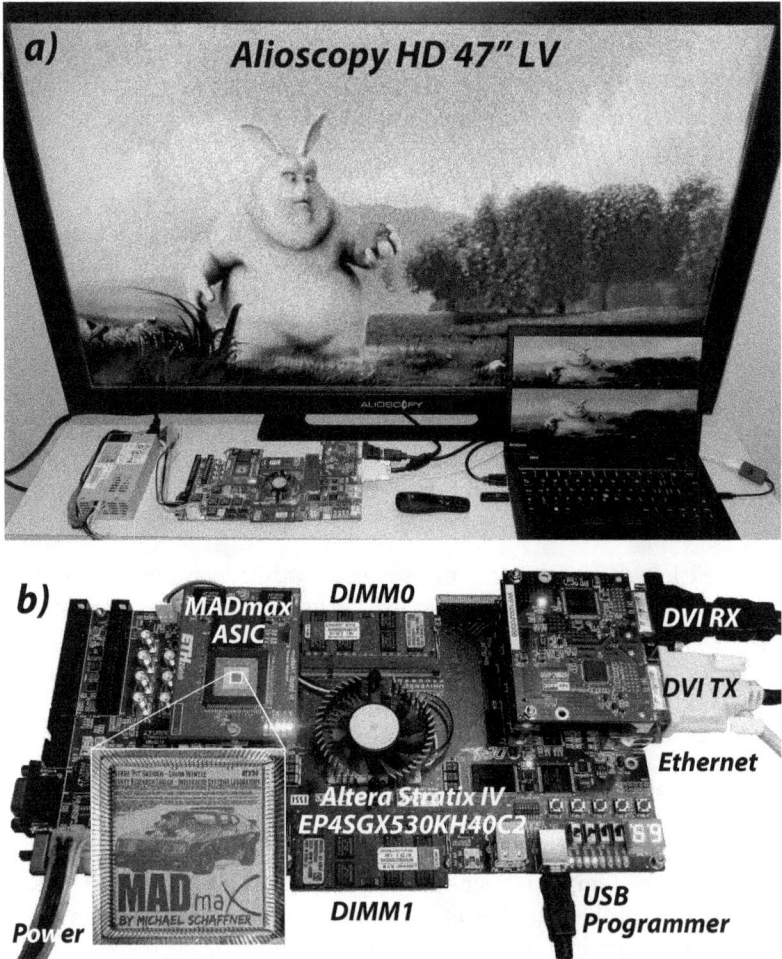

FIGURE 7.15: a) photograph of the running system, and b) close up of the hardware prototype, which can be mounted as an external display with video format 1920x2160@30Hz on any PC or laptop with DVI or HDMI capability. The MAD used in this experiment is an ALIOSCOPY HD 47" long-view display. Images copyright © 2008, BLENDER FOUNDATION, www.bigbuckbunny.org.

Chapter 8

Conclusions

In this thesis, we have investigated efficient algorithms and hardware architectures for a variety of subproblems arising in the context of IDW-based MVS. Experience collected in the domains of feature extraction, linear solvers and image resampling techniques for MADs culminated in the realization of a hardware demonstrator system for real-time MVS. The hardware IPs developed have been used to estimate circuit complexity and energy-efficiency in modern CMOS nodes. The system serves as a proof of concept, paving the way towards completely integrated solutions, e.g., as part of an SoC with possible deployment in TV sets, set-top boxes or mobile devices.

Subsequent work on STEA filtering resulted in a new method for the approximation of image-based regularization problems in a temporally consistent manner. The STEA filtering pipeline developed can be used to calculate accurate optical flow competing with related state-of-the-art methods in terms of speed and quality. The novel STEA method has the potential to improve the quality and efficiency of image processing algorithms such as IDW-based MVS significantly, since it requires less resources than previous STEA methods.

Finally, we investigated the use of LNS as an energy efficient alternative for floating-point arithmetic for applications where fixed-point arithmetic does not suffice. This led to the development of a framework able to generate accurate and approximate LNU instances. By integrating these units into a multi-processor system, we show that significant energy-efficiency gains can be achieved for specific signal and image processing kernels.

This chapter provides a summary of the main results of this thesis, and identifies possible directions and opportunities for future work.

8.1   Summary of Main Results

Pre-filtering for Efficient Resampling. Apart from rendering new view-points, MVS for MADs also involves a spatial multiplexing step that corresponds to a resampling operation of the views onto a display specific sampling pattern. This sampling pattern is typically a *non-orthogonal* arrangement. We discuss the artifacts that arise in cases with insufficient band limiting in Chapter 5, and show that they are more distinct and noticeable than in the case of orthogonal sampling arrangements. This is due to the fact that aliased frequency components of the images are subject to a *directionality* change (i.e., sinusoids with different propagation direction).

Further, we have analyzed different resampling strategies to mitigate these artifacts, and our evaluations show large differences in computational complexity, memory requirements and visual quality among the evaluated configurations. The highest visual quality was achieved by *post-filtering* algorithms that first resample each view separately, followed by a display-specific subsampling filter. However, this approach is computationally expensive, requires a lot of memory, and does not scale well with the number of views.

To this end we introduce a new *pre-filtering* concept where the anti-aliasing for the display sampling lattice is performed on the input images, before actually applying the warp transformation. This is possible, since the image deformations due to warping are relatively small and localized in the case of MVS. Algorithm variants employing this concept offer the best trade off between image quality and computational complexity – regardless of the input resolution. They also show better scalability with respect to the number of views than post-filtering approaches, since the anti-aliasing is performed on the few input images ($N < M$). In addition, only the effective display pixels have to be computed during warping.

Binary Descriptors Enable Real-Time Operation. For the applications at hand, *binary* descriptors work just as well in terms of recognition performance as non-binary descriptors like SIFT or SURF. In fact, our evaluations show that the SKB descriptor is an excellent choice for rectified stereo settings, where it outperforms other state-of-the-art descriptors (Section 2.3.4). Such binary descriptors are key for efficient hardware implementations, since they have a small memory footprint and typically require integer arithmetic computations only. In addition, their binary nature allows to efficiently calculate matching costs using the Hamming distance. Our hardware feature extraction and matching system presented in Chapter 2 bases on SKB, and distinguishes itself by a very high throughput of 15 k-25 k descriptors per frame, at 42 fps. This is an important property for MVS and disparity mapping applications relying on accurate sparse depth information. Related FPGA systems typically achieve only a throughput in the order of 512-2 k descriptors per frame.

Energy-Efficient IDW Solvers. As shown in Chapter 4, direct Cholesky decomposition solvers are the preferred solution for warp resolutions between 100×180 and 256×128 vertices (18 k-128 k variables)[1]in silicon-area-constrained settings. In addition, direct solvers exhibit constant runtime, which is a desirable property for real-time applications. The drawbacks of direct solvers are the size of the matrix decomposition, and the limited opportunities for parallelization. Scalability is thus limited, and pure on-chip embodiments are typically infeasible due to the size of the matrix decomposition.

As opposed to this, our estimations show that iterative conjugate gradient solvers offer the best energy efficiency tradeoff in scenarios where enough silicon area is available. While iterative solvers do not provide sufficient throughput and good energy efficiency when used with typical external memory interfaces present in today's mobile devices, this picture changes if on-chip SRAM is used exclusively. As opposed to direct solvers, such scenarios are feasible since memory requirements scale linearly with the matrix size. E.g., with a complexity of around $16 \, \text{mm}^2$, a conjugate gradient solver for 128 k variables is feasible in 28 nm CMOS technology, considering that complete SoCs typically have complexities beyond $100\,\text{mm}^2$.   Nevertheless, a large fraction of the area ($>90\,\%$) will be occupied by on-chip SRAM. Therefore, new memory technologies such as 2.5D integration or 3D stacking [157] have the potential to make such on-chip solvers possible. For example, emerging FPGA devices like the UltraScale+ series from XILINX [158] will contain significantly more on-chip SRAM than previous generations[2].

The First Hardware MVS Prototype.   We have designed and implemented a complete, IDW-based MVS system in hardware, that is able to synthesize high quality, full-HD content with up to eight views in real-time. During informal subjective tests using full-length S3D movie footage, we found that the system works well on a broad range of synthetic and live action content with the same parameters.

This system is the first of its kind.   The only other published complete system comprising similar functionality in terms of video analysis and MVS has been implemented on a high-end workstation. When comparing subcomponents of our system to related DIBR-based modules, we observe that some of the DIBR components are more compact in terms of logic and SRAM area. One of the main reasons for this is the fact that several related DIBR components assume perfectly rectified S3D input, which allows to reduce several 2D problems to 1D problems that are easier to implement. Our system and its components have been designed to tolerate small Keystone distortions, and therefore operate on 2D problems.

The current test bed is designed using an FPGA platform combined with a custom rendering ASIC, and our results show that a monolithic integration of the developed hardware IP into a SoC fabricated in 28 nm is completely feasible, and would amount to around $8.5\,\text{mm}^2$. With an estimated power consumption of ~1.3 W the accelerator would enable portable and energy efficient MVS, both essential properties for consumer electronic devices[3].

---

1. Although our evaluation includes smaller grids down to 128×64, lower resolutions than 100×180 may result in insufficient MVS quality and should not be used.
2. Announced capacities range up to 500 Mbit. For comparison, the second largest ALTERA Stratix IV device used in the MVS prototype has only 20 Mbit.
3. For comparison, the GPU subsystems of APPLE's A7 and NVIDIA's Tegra K1 occupy ~21 mm^2 and ~40 mm^2, respectively [22], and the power consumption (with display) of an A7-based tablet is ~6 W when running a graphics benchmark [247].

Commercial Feasibility of IDW-Based MVS. While previous work [33] found that algorithmic limitations (warping artifacts) will probably prevent commercial use of IDW for certain applications like aspect ratio retargeting, we think that this is not the case for MVS and depth remapping applications. While it is true that IDW has limitations and fails at rendering arbitrary new viewpoints (as needed in virtual reality applications and some 3DTV scenarios), IDW works well for small changes of the camera perspective. This can be attributed to the fact that the image texture is transformed in its entirety, without the need to fill disocclusion holes with inpainting. Since warping artifacts are subtle for small perspective changes, we conclude that IDW remains a viable alternative to DIBR techniques for applications with limited view transformations.

A New Binary Descriptor for Fast NNF Calculation. While SKB clearly outperforms state-of-the-art descriptors in rectified stereo settings, we realized that its recognition performance drops when used in settings with larger camera perspective changes (Section 3.5). To this end, we design a new binary descriptor termed BOOM, which outperforms other state of-the-art descriptors (binary and non-binary). Note, that neither of the two descriptors (SKB and BOOM) is better than the other in all cases. Rather, the two descriptors *complement* each other (Appendix 3.A). We employ BOOM to accelerate a recent NNF method for the calculation of quasi-dense, optical flow vector fields (Section 3.4). At the same levels of matching accuracy, our method is faster than related methods and has a leaner memory footprint due to the binary BOOM descriptor. Such NNFs are a crucial ingredient for many image-based applications like optical flow estimation and IDW. Our improved method paves the way towards real-time embodiments in hardware or mobile GPU implementations.

Fast STEA Filtering Pipeline. In Chapter 3, we have designed a STEA filtering method to approximate challenging regularization problems such as optical flow estimation, and to temporally stabilize noisy feature maps (such as visual saliency) that were calculated on a per frame basis. Our pipeline is more efficient than previous methods, since the temporal IIR formulation of the filter does not need to keep large temporal neighbourhoods in memory. Also, our pipeline does not depend on precomputed optical flow from an external method, as it is able to bootstrap its own accurate optical flow. This is achieved by using the filter developed to interpolate a quasi-dense NNF obtained with the new BOOM-based matching method[4]. Compared to state-of-the-art methods, our STEA method has lower computational cost and constant memory requirements, and is therefore suitable for real-time GPU and HW implementations. E.g., it could be used to enhance temporal consistency and scalability of real-time video processing systems such as the MVS prototype.

LNS for Non-linear Function Kernels. We have presented a framework to generate a family of accurate and approximate LNUs that are more compact than related state-of-the-art designs, while providing significantly more functionality. The similarity of LNUs with SFUs typically present in today's GPUs allows to conveniently add support for useful transcendental functions (2^x, $\log_2(x)$, $\sin(x)$, $\cos(x)$, $\text{atan2}(y,x)$). While the area cost of a single LNU is difficult to amortize in a single-core processing system, we show that an LNU *shared* among several processor cores is a viable alternative. Due to the fact that MUL, DIV, SQRT instructions can be efficiently computed in the integer ALUs of the cores, it is sufficient to share one LNU among four cores. Despite the fact that additions are more complex and energy-consuming in LNS, we show that the shared LNU designs can compute typical signal processing kernels up to 4.1× more energy-efficiently than an equivalent chip with four private FPUs. Further, we demonstrate that by relaxing the precision requirements from exact to 16 ULP, significant area savings of 40% can be achieved, and speedups can be further increased up to 5.54× in the best case due to lower LNU latency. Using examples from the image and audio processing domains, we show that the incurred quality losses are imperceptible in the evaluated precision range.

Given the current trend to incorporate more and more challenging signal processing algorithms into embedded and mobile systems, support for energy-efficient HDR arithmetic becomes ever more important – not only as enabler but also for convenience and rapid application

---

4. In fact, refinement of sparse optical flow in order to efficiently obtain a dense vector field is an emerging concept employed by several recent methods [91, 106, 108, 127].

development. Especially for applications involving frequent evaluation of complex non-linear functions, we believe that LNS arithmetic is a strong contender for HDR arithmetic.

8.2 General Observations on Efficiency

Approximations are Important for Efficient Circuits. This observation can be made in several application contexts. However, computer vision and graphics applications often allow for relatively coarse approximations [162] due to the fact that input data is often noisy to some extent, and outputs are interpreted by the human visual system that is inherently error tolerant. Approximations on the algorithmic, as well as on the micro-architectural and circuit levels can therefore be leveraged to enable efficient implementations, and can be found all over the place in this work. For example:

- The most popular technique on the micro-architectural level is custom arithmetic, such as fixed-point representations. If higher dynamic ranges are required, the use of custom floating-point formats is also an option. In particular, these formats need not be compliant with the IEEE standard and can therefore be tailored to the applications and implemented more efficiently. In addition, *operator fusing* is an important concept that can be leveraged to significantly reduce the complexity of floating-point datapaths. For example, the scalar product in the direct solver of the MVS prototype (Section 7.4.4) has been optimized in this way. For some function kernels containing many multiplications, divisions and square-roots, different number systems such as the LNS are also a viable option (Chapter 6).
- The binary features developed and employed in Chapters 2 and 3 produce binary vectors instead of numeric ones, and can therefore also be seen as an approximation. Moreover, such features are often computed using integral images and box filters that approximate Gaussian filters.
- Reduced filter orders and different filter arrangements such as the pre-filtering strategy for MADs presented in Chapter 5 represent another sort of approximations at the algorithmic level.

- Linear solvers for LS systems inherently comprise several approximation tradeoffs, such as the choice of suitable iteration and preconditioner combinations for example. Also, incomplete factorizations, or a reduced amount of iterations can be used to compute an approximate solution (Chapter 4).
- STEA filtering can be seen as an approximation in the sense that it can be employed to tackle regularization problems in an alternative way than just by solving the LS equation systems using standard methods (Chapter 3).

While some of these points may seem obvious, they are nevertheless important, and should be carefully evaluated to make sure that application quality targets are still met. Unfortunately it is possible that numerical metrics (such as PSNR, AEE, ROC and the like) do not correctly capture the artifacts. As a consequence, this may lead to suboptimal design decisions. Therefore evaluations must also include subjective assessments in addition to numerical metrics.

Reducing Data Movement is Key. Other design patterns gaining increased importance in modern CMOS technologies is to increase data locality and reduce data movement. Keeping the working data close to the processing elements by means of local memories and caches has been an established concept for many years. But its primary goal was to decrease *access latency* [248]. In modern CMOS technologies, the discrepancy between the energy needed for a compute operation on-chip and an off-chip memory access is as high as never before [249]: An off-chip memory access costs orders of magnitudes more than a compute operation on-chip. We can observe the effects of these developments by looking at the 28 nm energy efficiency estimates of the MVS prototype (Section 7.5.3), where more than 50 % of the energy is dissipated just to move data to and from DRAM. This has consequences for the design of energy efficient hardware accelerators. Algorithmic rearrangements and approximations facilitating on-the-fly computations and streaming operation should be performed, where possible. If enough silicon area is available[5], keeping the data on-chip can lead to significant increases in terms of

---

5. This is to some extent an economic question. However, using large on-chip memories can counteract dark silicon limitations to some degree [250].

energy-efficiency (see also solver evaluations in Chapter 4). To this end, near-memory or in-memory computing [249, 251] may prove to be a vital concept for future compute architectures.

8.3 Outlook and Future Work

Several aspects of the IDW pipeline could be investigated further. The most important ones are sketched below.

Improved Solvers for Next Generation MVS. Although we found that a warp resolution of 180×100 is sufficient for MVS for current full-HD MADs, higher MAD resolutions in the order of 4 k or 8 k combined with better optics and display fidelity may dictate higher warp resolutions. This creates a need for more efficient solvers than analyzed in this work.

Direct solvers exhibit many data-dependencies and therefore offer only limited parallelization opportunities. In addition, the memory required for the matrix decomposition grows super-linearly with the number of variables. Therefore, iterative solvers are the way to go when problems are scaled up, since the operations can be parallelized better, and the storage requirements grow only linearly with the amount of variables to be solved for. On the algorithmic level, application-specific, parallel preconditioners or multigrid schemes [133, 156, 161] should be investigated in order to ensure quick convergence and to keep memory bandwidth within acceptable bounds. From the technology side, new memory technologies should be evaluated allowing the solver to keep all data completely on-chip.

Note however, that pixel-dense warps will most likely never be needed, even for higher MAD resolutions (Section 7.5.1). Moderately increased warp resolutions, combined with better features such as quasi-dense NNFs (Section 3.C) may be the best way to align MVS quality with next generation display systems.

Alternative Approaches to Compute Warps. STEA filtering is a viable alternative to solvers for image-based regularization problems like optical flow- or depth estimation. Especially when incorporating the temporal dimension, STEA proves to scale better with the amount of pixels than solver-based approaches, and could therefore also prove useful for computing image domain warps. However, since STEA filtering only computes an *approximation*, further work including subjective studies is needed to assess the feasibility for IDW in terms of image quality. Apart from STEA, there are two other alternative ways to compute image warps that may be worth to be analyzed in more detail:

- *Finite element image warping* with adaptive meshes could be used instead of the regular, finite difference mesh employed in this work. This technique can reduce the amount of variables to be solved for by adaptively reducing the degrees of freedom of the mesh, based on the detail level of the input images. This technique has recently been shown to work well in the context of retargeting [252] and MVS [253], and should be investigated closer in order to see whether this leads to more efficient hardware architectures. These methods are conceptually more involved than finite difference methods, since the warp mesh is not regular anymore.

- Another warp approximation could be obtained by *training a convolutional neural network (CNN)* to calculate warping functions. Similar approaches have been shown to work well in the context of sparse-to-dense conversion of optical flow vectors [106, 107]. CNNs may prove viable for warp calculation from a computational viewpoint, since the warp resolution is typically significantly smaller than the resolution of the images.

Finally, IDW could also be adapted to enable robust and automatic MVS for other display types such as light-field devices, as shown in a preliminary study by HOLOGRAPHIKA [254].

Approximate Computing.   Some algorithms like iterative solvers or machine-learning-based approaches are inherently robust against quantization noise. E.g., mixed precision concepts [142, 255] that perform the heavy lifting work using low-precision arithmetic, followed by a high-precision refinement iteration may be applicable to IDW problems too. A similar concept, based on approximate LNS operators (so-called low precision high dynamic range (LPHDR) arithmetic) is used in [256,257] to efficiently implement a feature tracking algorithm.

Apart from traditional word width reductions, novel approximate computing concepts that also take into account probabilistic effects of the implementation technology [162,258] could be evaluated in this context. Memory compression techniques could be combined with these ideas, since certain number representations may turn out to be well compressible [259].

Further Steps Towards SoC Integration. To move further towards a commercial embodiment in the form of an SoC, the system specs should first be aligned with next-generation MADs (higher resolution and number of views). Some of the suggested improvements, like improved NNF features, higher warp resolutions and different resampling strategies for MADs may be implemented in this step as well. A next milestone before actual integration in silicon should then be a full-system emulation using a heterogeneous FPGA platform, such as the new Zynq MPSoC Series from XILINX [260] that also includes an ARM application processor and a mobile GPU. The fixed function IPs could then be implemented in the FPGA fabric, and the application processor could be used to perform the control tasks or perform computationally less heavy operations that are likely to be updated frequently as part of a firmware upgrade (e.g., setting up the IDW matrices). If a mobile GPU is available, using it to implement the LS solver could be an interesting design alternative. The solver is the only part of the MVS system requiring floating-point arithmetic, and might get algorithmically more complex if higher warp-resolutions, or adaptive warping meshes are employed. Implementing the solver on the GPU would reduce the design effort, which may be substantial when switching to a more complex solver algorithm.

Appendix A

Chip Gallery

This appendix lists all chips that are related to the present thesis. For a complete list, see also http://asic.ethz.ch/authors/Michael_ Schaffner.html.

A.1   Esper

Nonlinear image warping or image resampling is a necessary step in many current and upcoming video applications, such as video retargeting, S3D mapping, and MVS. We employ an EWA splatting approach to 2-D image resampling, and extend the classical EWA framework for increased visual quality. The hardware architecture designed is able to render high-quality video sequences in real time, and the resulting chip is targeted for low-power applications in end-user display devices.

Designers	Michael Schaffner, Marian Runo, Pierre Greisen, Simon Heinzle, Frank K. Gürkaynak
Application / Publications	Graphics / [51, 52]
Technology / Package	UMC 180 nm/ PGA120
Dimensions / Gates	3240 µm ×3240 µm/ 700 kGE
Voltage / Power	1.8V / 200 mW
Clock	125 MHz

A.2 Vesper

This design is a continuation of Esper, and the main focus of this work was to extend the design to support full HD and to add interfaces to commercially available components such as QDRII SDRAM and DVI. Two clock domains decouple the high-performance rendering part from the fixed rate DVI interface.

Designers	Richard Emler, Pierre Greisen, Simon Heinzle, Michael Schaffner, Marian Runo, Patrice Guillet, Frank K. Gürkaynak
Application / Publications	Graphics / [50–52]
Technology / Package	UMC 130 nm/ PGA256
Dimensions / Gates	5000 µm ×5000 µm/ 1800 kGE
Voltage / Power	1.2V
Clock	180 MHz

A.3   EvaLongoria

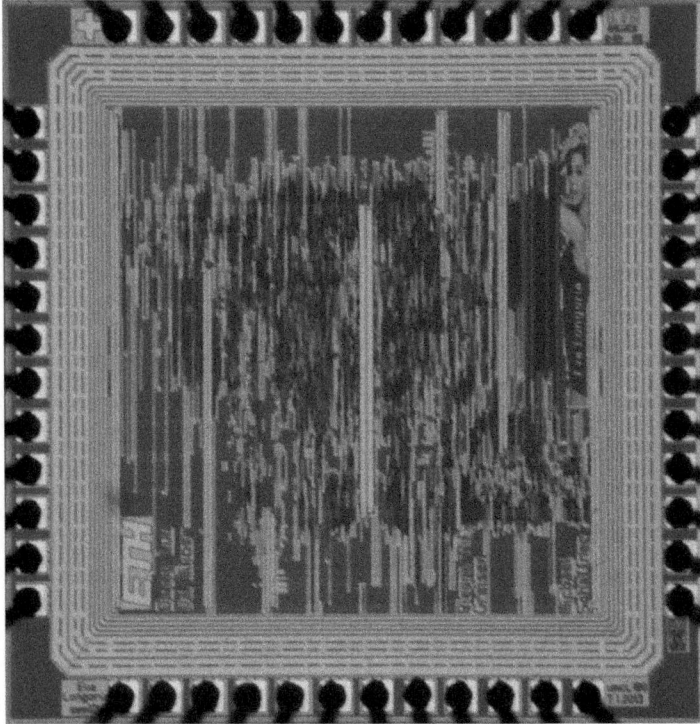

This is an alternative resampling chip for IDW that employs bicubic interpolation instead of EWA splatting.

Designers	Val Mikos, Danny Luu, Michael Schaffner Pierre Greisen
Application / Publications	Graphics / [52]
Technology / Package	UMC 180 nm/ QFN56
Dimensions / Gates	1525 µm ×1525 µm/ 120 kGE
Voltage / Power	1.8V / 140 mW
Clock	140 MHz

A.4 Madmax

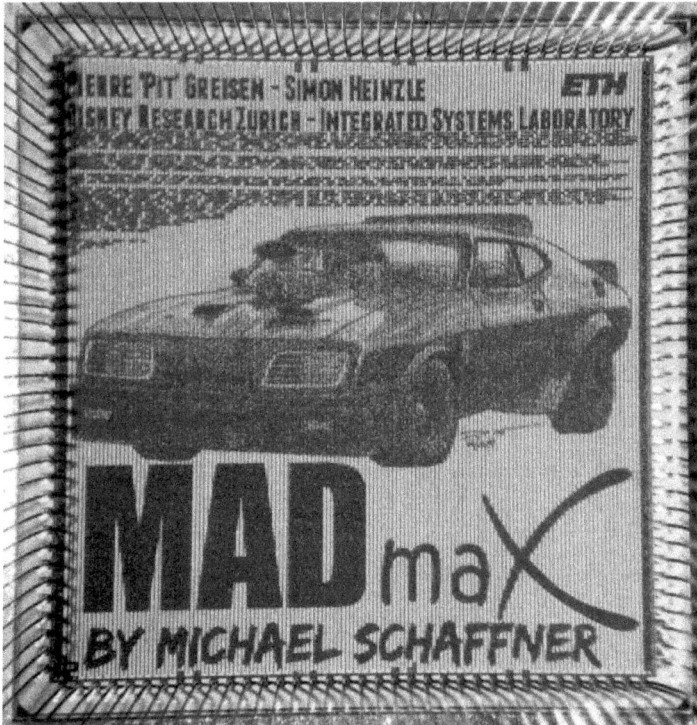

This rendering core is a continuation of the Esper and Vesper chips, and extends EWA splatting to MVS. The chip takes S3D content plus warping grids as input and produces high-quality, alias-free, multiview output ready to be viewed on a MAD.

Designers	Michael Schaffner, Pierre Greisen, Simon Heinzle, Frank K. Gürkaynak
Application / Publications	Graphics / [36]
Technology / Package	UMC 65 nm/ CQFP120
Dimensions / Gates	3772 µm ×3772 µm/ 7000 kGE
Voltage / Power	1.2V / 750 mW
Clock	400 MHz

A.5 Sandstorm

This ASIC implements the interest point detection and descriptor extraction stages of the SKB method for finding point-correspondences in S3D video streams.

Designers	Lukas Cavigelli, Pascal Alexander Hager, Oliver With, Michael Schaffner, Pierre Greisen
Application / Publications	Vision / [42, 44]
Technology / Package	UMC 180 nm/ CQFP120
Dimensions / Gates	3240 µm × 1525 µm/ 250 kGE
Voltage / Power	1.8V / 146 mW
Clock	100 MHz

A.6 Artemis

This is one of a series of chips that is part of the PULP project which add various floating point units to a cluster of four processors. Artemis is the base version which contains a private IEEE compliant FPU for each core.

Designers	Lukas Mueller, Thomas Gautschi, Michael Gautschi, Michael Schaffner, Frank K. Gürkaynak
Application / Publications	PULP [38–40, 53]
Technology / Package	UMC 65 nm/ QFN40
Dimensions / Gates	1252 µm ×1252 µm/ 600 kGE
Voltage / Power	1.2V / 120 mW
Clock	500 MHz

A.7   Hecate

This is one of a series of chips that is part of the PULP project which add various floating point units to a cluster of four processors. Hecate contains two IEEE compliant FPUs that are shared among all cores.

Designers	Youri Popoff, Florian Scheidegger, Michael Gautschi, Michael Schaffner, Frank K. Gürkaynak
Application / Publications	PULP [38–40, 53]
Technology / Package	UMC 65 nm/ QFN40
Dimensions / Gates	1252 µm ×1252 µm/ 600 kGE
Voltage / Power	1.2V / 120 mW
Clock	500 MHz

A.8 Selene

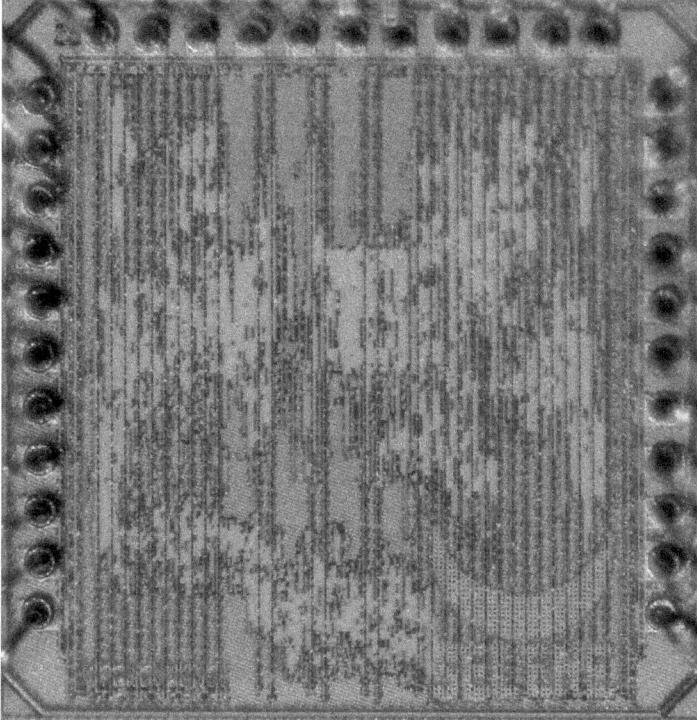

This is one of a series of chips that is part of the PULP project which add various floating point units to a cluster of four processors. Selene contains one LNU that is shared among all cores.

Designers	Youri Popoff, Florian Scheidegger, Michael Gautschi, Michael Schaffner, Frank K. Gürkaynak
Application / Publications	PULP [38–40, 53]
Technology / Package	UMC 65 nm/ QFN40
Dimensions / Gates	1252 µm ×1252 µm/ 600 kGE
Voltage / Power	1.2 V/ 120 mW
Clock	500 MHz

A.9   Phoebe

This is one of a series of chips that is part of the PULP project which add various floating point units to a cluster of four processors. Phoebe contains an improved and extended LNU that is shared among all cores.

Designers	Michael Gautschi, Michael Schaffner
Application / Publications	PULP [38–40, 53]
Technology / Package	UMC 65 nm/ QFN40
Dimensions / Gates	1252 µm ×1252 µm/ 600 kGE
Voltage / Power	1.2 V/ 120 mW
Clock	500 MHz

A.10 Hydra

Hydra is an edge-aware filtering chip that uses a tiled version of the PF. I has been specifically designed for spreading sparse data, and is able to filter 720p content at 30 fps.

Designers	Christelle Gloor, Manuel Eggimann, Lukas Cavigelli, Michael Schaffner
Application / Publications	Vision / [49]
Technology / Package	UMC 65 nm/ QFN64
Dimensions / Gates	1875 µm ×1875 µm/ 1400 kGE
Voltage / Power	1.2 V/ 800 mW
Clock	333 MHz

A.11 Phoenix

This design is an adaption of the Sandstorm ASIC and has been implemented to compare the hardware complexity of the SKB and BOOM descriptors. It implements the same interest point detector used for SKB, combined with a BOOM descriptor extraction stage.

Designers	Bastien Joye, Yannick Ruffiner, Pascal Hager, Michael Schaffner
Application / Publications	Vision / [49]
Technology / Package	UMC 65 nm/ QFN40
Dimensions / Gates	1252 µm ×1252 µm/ 600 kGE
Voltage / Power	1.2 V/ ? mW
Clock	300 MHz

Appendix B

Notation and Acronyms

Symbols

$\mathbf{A}$	matrix
$\mathbf{a}$	vector
a, A	scalars
$\mathbb{N}$	set of natural numbers
$\mathbb{Z}$	set of integer numbers
$\mathbb{R}$	set of real numbers
$\mathbb{R}^n$	set of real n vectors
$\mathbb{R}^{n \times n}$	set of real $n \times n$ matrices
$\mathcal{D}$	domain for coordinates or iterators in $\mathbb{R}^n$ or $\mathbb{Z}^n$
Λ	sampling lattices/grids
$E(.)$	energy functionals
λ	bias parameter
ϵ	numerical error

Operators

$(\cdot)^T$	matrix transposition

$(\cdot)^{-1}$ inverse function, inverse of square matrix
$|\cdot|$ absolute value
$\lceil\cdot\rceil$ ceil: smallest integer value equal to or larger as argument
$\lfloor\cdot\rfloor$ floor: largest integer value equal to or smaller as argument
$\|\cdot\|_1$ ℓ^1-norm
$\|\cdot\|$ ℓ^2-norm or Euclidean norm
$\|\cdot\|_\infty$ ℓ^∞-norm
$\|\cdot\|_p$ ℓ^p-norm
$\log_2$ base-2 logarithm
$\log_{10}$ base-10 logarithm
$\mathrm{mod}(\cdot)$ modulo operator
Σ summation
Π product sequence
$\mathcal{O}(\cdot)$ big-oh asymptotic bound

Acronyms

AEE   average endpoint error
ALU   arithmetic logic unit
ASIC   application-specific integrated circuit
ASIP   application-specific instruction-set processor
ATE   area time energy
AT   area time
AVC   advanced video coding
AVX   advanced vector extension

BF   bilateral filter
BHOG   binarized histogram of oriented gradients
BiCGStab   biconjugate gradient stabilized (Section 4.3.3)
BLAS   basic linear algebra subprograms
BOOM   binarized octal orientation maps
BRIEF   binary robust independent elementary features
BRISK   binary rotation invariant scalable features

CG   conjugate gradients (Section 4.3.3)
CGLS   conjugate gradients least-squares (Section 4.3.3)
CHD   Cholesky decomposition (Section 4.3.3)

CLAHE	contrast-limited adaptive histogram equalization
CMOS	complementary metal-oxide semiconductor
CNN	convolutional neural network
CODEC	coder/decoder
CORDIC	coordinate rotation digital computer
CPM	coarse-to-fine patch match
CPU	central processing unit
DAISY	(a fast local descriptor for dense matching)
DCT	discrete cosine transform
DCU	descriptor calculation unit
DDR	double data rate
DIBR	depth image based rendering
DIMM	dual inline memory module
DOB	difference of boxes
DRAM	dynamic random access memory
DSP	digital signal processing
DT	domain transform
DVI	digital visual interface
EA	edge-aware
EWA	elliptical weighted average
FAST	features from accelerated segment test
FDR	false discovery rate
FE	full-adder equivalents
FFT	fast Fourier transform
FIFO	first-in first-out
FIR	finite impulse response
FLANN	fast library for approximate nearest neighbors
FLOP	floating point operation
FP	floating point
FPGA	field-programmable gate array
FPR	false-positive rate
FPU	floating point unit
FREAK	fast retina keypoint
FTV	free-viewpoint television

GE   gate equivalents
GF   guided filter
GPU   graphics processing unit

HD   high-definition
HDMI   high-definition multimedia interface
HDR   high dynamic range
HEVC   high-efficiency video coding
HOG   histogram of oriented gradients
HPC   high-performance computing
HW   hardware

ICHOL   incomplete Cholesky decomposition (Section 4.3.3)
IDW   image-domain warping
IIR   infinite impulse response
IoT   internet of things
IP   intellectual property
IPC   instructions per cycle
ISP   image signal processor

LDA   linear discriminant analysis
LDDB   learning-based local difference binary
LL   low leakage
LLVM   low level virtual machine
LNS   logarithmic number system
LNU   logarithmic number unit
LOG   Laplacian of Gaussian
LPDDR   low-power double data rate
LPHDR   low precision high dynamic range
LS   least-squares
VPU   vision processing unit
LSB   least significant bit
LSQR   least-squares QR (Section 4.3.3)
LUT   lookup table

MAC   multiply-accumulate
MAD   multiview autostereoscopic display
MINRES   minimum residual (Section 4.3.3)

MPEG	moving picture expert group
MSB	most significant bit
MV	multiview
MVS	multiview synthesis
NMS	non-maximum suppression
NN	nearest neighbour
NNF	nearest neighbour field
PC	personal computer
PCA	principal component analysis
PCB	printed circuit board
PCG	preconditioned conjugate gradients (Section 4.3.3)
PCS	partial carry save
PF	permeability filter
PHY	physical layer
PSNR	peak signal-to-noise ratio
PTV	process temperature voltage
PULP	parallel ultra-low power
QFT	quaternion Fourier transform
RAM	random access memory
RANSAC	random sample consensus
RGB	red green blue
RISC	reduced instruction set computer
ROC	receiver operating characteristic
ROM	read-only memory
RTL	register transfer level
RX	receiver
S3D	stereoscopic 3D
SAIF	switching activity interchange format
SF	special function
SFE	special function extension
SFU	special function unit
SIFT	scale-invariant feature transform
SIMD	single instruction multiple data

SKB	semantic kernels binarized
SLAM	simultaneous localization and mapping
SoC	system on chip
SOS	second order section
SRAM	static random access memory
SSE	streaming SIMD extensions
STEA	spatio-temporal edge-aware
SURF	speeded-up robust features
SVD	singular value decomposition (Section 4.3.3)
SVM	support vector machine
SW	software

THD+N	total harmonic distortion plus noise
TPR	true-positive rate
TV	television
TX	transmitter

| ULP | unit in the last place |
| USB | universal serial bus |

| VCD | value change dump |

| WLS | weighted least squares |

Bibliography

[1] B. Mendiburu, "3D Movie Making," *Focal Express*, 2009.

[2] A. Smolic, P. Kauff, S. Knorr, A. Hornung *et al.*, "Three-Dimensional Video Postproduction and Processing," *PIEEE*, vol. 99, no. 4, pp. 607–625, 2011.

[3] C. Zhu, Y. Zhao, L. Yu, and M. Tanimoto, *3D-TV System with Depth-Image-Based Rendering*. Springer, 2014.

[4] N. A. Dodgson, "Optical Devices: 3D without the Glasses," *Nature*, vol. 495, no. 7441, pp. 316–317, March 2013.

[5] J. Konrad and M. Halle, "3-D Displays and Signal Processing," *IEEE SPM*, vol. 24, no. 6, pp. 97–111, 2007.

[6] A. Boev, R. Bregovic, and A. Gotchev, "Signal Processing for Stereoscopic and Multi-View 3D Displays," in *Handbook of Signal Processing Systems*. Springer New York, 2013, pp. 3–47.

[7] D. Tian, P. Lai, P. Lopez, and C. Gomila, "View Synthesis Techniques for 3D Video," *Proc. SPIE*, vol. 7443, p. 74430T, 2009.

[8] A. Smolic, S. Poulakos, S. Heinzle, P. Greisen *et al.*, "Disparity-Aware Stereo 3D Production Tools," in *IEEE CVMP*, 2011, pp. 165–173.

[9] M. Tanimoto, M. Tehrani, T. Fujii, and T. Yendo, "Free-Viewpoint TV," *IEEE SPM*, vol. 28, no. 1, pp. 67–76, Jan 2011.

[10] L. Onural, T. Sikora, J. Ostermann, A. Smolic *et al.*, "An Assessment of 3DTV Technologies," in *NAB BEITC*, 2006, pp. 456–467.

[11] N. S. Holliman, N. A. Dodgson, G. E. Favalora, and L. Pockett, "Three-Dimensional Displays: a Review and Applications Analysis," *IEEE TBC*, vol. 57, no. 2, pp. 362–371, 2011.

[12] S. Reichelt, R. Häussler, G. Fütterer, and N. Leister, "Depth Cues in Human Visual Perception and Their Realization in 3D Displays," in *SPIE Defense, Security, and Sensing*, 2010, pp. 76 900B–76 900B.

[13] M. Tanimoto, T. Fujii, and K. Suzuki, "View synthesis algorithm in view synthesis reference software 3.5 (vsrs3.5)," ISO/IEC JTC1/SC29/WG11 (MPEG, Tech. Rep. M16090, May 2009.

[14] N. Stefanoski, O. Wang, M. Lang, P. Greisen *et al.*, "Automatic View Synthesis by Image-Domain-Warping," *IEEE TIP*, vol. 22, no. 9, pp. 3329–3341, 2013.

[15] G. Wolberg, "Digital Image Warping," *IEEE Computer Society Press*, vol. 3, 1990.

[16] P. Ndjiki-Nya, M. Köppel, D. Doshkov, H. Lakshman *et al.*, "Depth Image Based Rendering with Advanced Texture Synthesis," in *IEEE ICME*, July 2010.

[17] R. T. Pushpalwar and S. H. Bhandari, "Image Inpainting Approaches – A Review," in *IEEE IACC*, Feb 2016, pp. 340–345.

[18] P. Patel, A. Prajapati, and S. Mishra, "Review of Different Inpainting Algorithms," *IJCA*, vol. 59, no. 18, 2012.

[19] Samsung, "Exynos," Accessed Jan 2017. [Online]. Available: http://www.samsung.com/semiconductor/minisite/Exynos

[20] A. Skende, "Introducing 'Parker' Next-Generation Tegra System-On-Chip," 2016, Hot Chips, Accessed Jan 2017. [Online]. Available: http://www.hotchips.org/wp-content/uploads/hc_archives/ hc28/HC28.22-Monday-Epub/HC28.22.30-Low-Power-Epub/ HC28.22.322-Tegra-Parker-AndiSkende-NVIDIA-v01.pdf

[21] Qualcomm, "Snapdragon Processors and Modems," Accessed Jan 2017. [Online]. Available: http://www.qualcomm.com/products/snapdragon

[22] A. Eassa, "Just How Big Is NVIDIA Corporation's Tegra K1?" June 2014, Accessed Jan 2017. [Online]. Available: http://www.fool.com/investing/general/2014/06/04/just-how-big-is-nvidia-corporations-tegra-k1.aspx

[23] Gartner, Inc., "SOCs Are a Major, Multifaceted Semiconductor Theme," 2005, Accessed Jan 2017. [Online]. Available: http://www.gartner.com/doc/486402/socs-major-multifaceted-semiconductor-theme

[24] Y.-K. Chen and S. Y. Kung, "Trend and Challenge on System-on-a-Chip Designs," *JSPS*, vol. 53, no. 1, pp. 217–229, 2008.

[25] M. Briggs and P. Zarkesh-Ha, "Evaluating Mobile SOCs as an Energy Efficient DSP Platform," in *IEEE SOCC*, Sept 2014, pp. 293–298.

[26] Movidius, "Vision Processing Unit," Accessed Jan 2017. [Online]. Available: http://www.movidius.com/solutions/vision-processing-unit

[27] H. Kaeslin, "Top-Down Digital VLSI Design, from VLSI Architectures to Gate-Level Circuits and FPGAs," *Morgan Kaufmann*, 2014.

[28] R. Hartley and A. Zisserman, *Multiple View Geometry in Computer Vision*. Cambridge University Press, 2003.

[29] P. Krähenbühl, M. Lang, A. Hornung, and M. Gross, "A System for Retargeting of Streaming Video," *ACM TOG*, vol. 28, no. 5, p. 126, 2009.

[30] M. Rubinstein, D. Gutierrez, O. Sorkine, and A. Shamir, "A Comparative Study of Image Retargeting," in *ACM TOG*, vol. 29, no. 6, 2010, p. 160.

[31] S. Kopf, B. Guthier, C. Hipp, J. Kiess *et al.*, "Warping-Based Video Retargeting for Stereoscopic Video," *IEEE ICIP*, 2014.

[32] M. Lang, A. Hornung, O. Wang, S. Poulakos *et al.*, "Nonlinear Disparity Mapping for Stereoscopic 3D," *ACM TOG*, vol. 29, no. 4, p. 75, 2010.

[33] P. Greisen, "Hardware architectures for real-time video processing and view synthesis," Ph.D. dissertation, ETH Zürich, Rämistrasse 101, 8092 Zürich, Switzerland, 2013, Diss. ETH No. 21116.

[34] L. Itti, C. Koch, and E. Niebur, "A Model of Saliency-based Visual Attention for Rapid Scene Analysis," *IEEE TPAMI*, vol. 20, 1998.

[35] C. Guo, Q. Ma, and L. Zhang, "Spatio-Temporal Saliency Detection Using Phase Spectrum of Quaternion Fourier Transform," in *IEEE CVPR*, June 2008, pp. 1–8.

[36] M. Schaffner, P. Greisen, S. Heinzle, F. K. Gürkaynak *et al.*, "MADmax: A 1080p Stereo-to-Multiview Rendering ASIC in 65 nm CMOS based on Image Domain Warping," in *ESSCIRC*, 2013, pp. 61–64.

[37] A. Chapiro, "Perceptual Enhancements for Novel Displays," Ph.D. dissertation, ETH Zürich, Rämistrasse 101, 8092 Zürich, Switzerland, 2015, Diss. ETH No. 23029.

[38] M. Gautschi, M. Schaffner, F. K. Gürkaynak, and L. Benini, "A 65nm CMOS 6.4-to-29.2pJ/FLOP@0.8V Shared Logarithmic Floating Point Unit for Acceleration of Nonlinear Function Kernels in a Tightly Coupled Processor Cluster," in *IEEE ISSCC*, 2016.

[39] M. Gautschi, M. Schaffner, F. K. Gürkaynak, and L. Benini, "An Extended Shared Logarithmic Unit for Nonlinear Function Kernel Acceleration in a 65-nm CMOS Multicore Cluster," *IEEE JSSC*, vol. 52, no. 1, pp. 98–112, Jan 2017.

[40] M. Schaffner, M. Gautschi, F. K. Gürkaynak, and L. Benini, "Accuracy and Performance Trade-offs of Logarithmic Number Units in Multi-Core Clusters," in *IEEE ARITH*, 2016.

[41] M. Schaffner, P. Greisen, S. Heinzle, and A. Smolic, "Efficient Image Resampling for Multiview Displays," in *IEEE ICASSP*, 2013.

[42] M. Schaffner, P. Hager, L. Cavigelli, P. Greisen *et al.*, "A Real-Time 720p Feature Extraction Core Based on Semantic Kernels Binarized," in *IFIP/IEEE VLSI-SoC*, Oct 2013, pp. 27–32.

[43] M. Schaffner, F. K. Gürkaynak, A. Smolic, H. Kaeslin *et al.*, "An Approximate Computing Technique for Reducing the Complexity of a Direct-Solver for Sparse Linear Systems in Real-Time Video Processing," in *DAC*, 2014.

[44] M. Schaffner, P. A. Hager, L. Cavigelli, Z. Fang *et al.*, "A Complete Real-Time Feature Extraction and Matching System Based on Semantic Kernels Binarized," in *VLSI-SoC: At the Crossroads of Emerging Trends.* Springer Berlin Heidelberg, 2015.

[45] M. Schaffner, F. K. Gürkaynak, A. Smolic, and L. Benini, "DRAM or no-DRAM? Exploring Linear Solver Architectures for Image Domain Warping in 28 nm CMOS," in *DATE*, 2015.

[46] M. Schaffner, F. K. Gürkaynak, H. Kaeslin, L. Benini *et al.*, "Automatic Multiview Synthesis – Towards a Mobile System on a Chip," in *IEEE VCIP*, Dec 2015, pp. 1–4.

[47] M. Schaffner, F. K. Gürkaynak, H. Kaeslin, L. Benini *et al.*, "Automatic Multiview Synthesis – Prototype Demo," in *IEEE VCIP)*, Dec 2015, pp. 1–1.

[48] M. Schaffner, F. K. Gürkaynak, P. Greisen, H. Kaeslin *et al.*, "Hybrid ASIC/FPGA System for Fully Automatic Stereo-to-Multiview Conversion Using IDW," *IEEE TCSVT*, vol. 26, no. 11, pp. 2093–2108, Nov 2016.

[49] M. Schaffner, F. Scheidegger, L. Cavigelli, H. Kaeslin *et al.*, "Towards Spatio-Temporal Edge-Aware Filtering in Real-Time," *IEEE TIP*, 2017, submitted.

[50] P. Greisen, R. Emler, M. Schaffner, S. Heinzle *et al.*, "A General-transformation EWA View Rendering Engine for 1080p Video in 130nm CMOS," in *IEEE/IFIP VLSI-SoC*, 2012, pp. 105–110.

[51] P. Greisen, M. Schaffner, S. Heinzle, M. Runo *et al.*, "Analysis and VLSI Implementation of EWA Rendering for Real-Time

BIBLIOGRAPHY

HD Video Applications," *IEEE TCSVT*, vol. 22, no. 11, pp. 1577–1589, Nov 2012.

[52] P. Greisen, M. Schaffner, D. Luu, V. Mikos *et al.*, "Spatially-Varying Image Warping: Evaluations and VLSI Implementations," in *VLSI-SoC: From Algorithms to Circuits and System-on-Chip Design.* Springer Berlin Heidelberg, 2013, vol. 418, pp. 64–87.

[53] Y. Popoff, F. Scheidegger, M. Schaffner, M. Gautschi *et al.*, "High-Efficiency Logarithmic Number Unit Design Based on an Improved Co-Transformation Scheme," in *DATE*, 2016.

[54] P. Niklaus, T. Birchler, T. Aebi, M. Schaffner *et al.*, "StoneNode: A Low-Power Sensor Device for Induced Rockfall Experiments," in *IEEE SAS*, 2017.

[55] D. Lowe, "Distinctive Image Features From Scale-Invariant Keypoints," *IJCV*, vol. 60, no. 2, pp. 91–110, 2004.

[56] H. Bay, T. Tuytelaars, and L. Van Gool, "SURF: Speeded up Robust Features," *ECCV*, pp. 404–417, 2006.

[57] M. Calonder, V. Lepetit, M. Ozuysal, T. Trzcinski *et al.*, "BRIEF: Computing a Local Binary Descriptor Very Fast," in *IEEE TPAMI*, vol. 34, no. 7, 2012, pp. 1281–1298.

[58] S. Leutenegger, M. Chli, and R. Siegwart, "BRISK: Binary Robust Invariant Scalable Keypoints," in *IEEE ICCV*, 2011, pp. 2548–2555.

[59] A. Alahi, R. Ortiz, and P. Vandergheynst, "FREAK: Fast Retina Keypoint," in *IEEE CVPR*, 2012, pp. 510–517.

[60] F. Zilly, C. Riechert, P. Eisert, and P. Kauff, "Semantic Kernels Binarized - A Feature Descriptor for Fast and Robust Matching," in *CVMP*, Nov 2011, pp. 39–48.

[61] D. Scharstein and R. Szeliski, "High-Accuracy Stereo Depth Maps Using Structured Light," in *IEEE CVPR*, vol. 1, 2003, pp. I–195.

[62] J. Park, H. Kim, and L. Kim, "A 182 mW 94.3 f/s in Full HD Pattern-Matching Based Image Recognition Accelerator for an Embedded Vision System in 0.13-CMOS Technology," *IEEE TCSVT*, vol. 23, no. 5, pp. 832–845, 2013.

[63] J. Wang, S. Zhong, L. Yan, and Z. Cao, "An Embedded System-on-Chip Architecture for Real-time Visual Detection and Matching," *IEEE TCSVT*, vol. 24, no. 3, pp. 525–538, March 2014.

[64] J. Vourvoulakis, J. Kalomiros, and J. Lygouras, "FPGA Accelerator for Real-Time SIFT Matching with RANSAC Support," *MICPRO*, vol. PP, 2016.

[65] P. Viola and M. Jones, "Rapid Object Detection using a Boosted Cascade of Simple Features," in *IEEE CVPR*, vol. 1, 2001, pp. I-511–I-518 vol.1.

[66] V. Bonato, E. Marques, and G. Constantinides, "A Parallel Hardware Architecture for Scale and Rotation Invariant Feature Detection," *IEEE TCSVT*, vol. 18, no. 12, pp. 1703–1712, 2008.

[67] J. Svab, T. Krajnik, J. Faigl, and L. Preucil, "FPGA Based Speeded Up Robust Features," in *TePRA*, 2009, pp. 35–41.

[68] D. Bouris, A. Nikitakis, and J. Walters, "Fast and Efficient FPGA-Based Feature Detection Employing the SURF Algorithm," in *IEEE FCCM*, 2010, pp. 3–10.

[69] M. Schaeferling and G. Kiefer, "Object Recognition on a Chip: A Complete SURF-Based System on a Single FPGA," in *ReConFig*, 2011, pp. 49–54.

[70] T. Sledevic and A. Serackis, "SURF Algorithm Implementation on FPGA," in *BEC*, 2012, pp. 291–294.

[71] D. Jeon, Y. Kim, I. Lee, Z. Zhang *et al.*, "A 470mV 2.7mW Feature Extraction-Accelerator for Micro-Autonomous Vehicle Navigation in 28 nm CMOS," in *ISSCC*, 2013, pp. 166–167.

[72] E. Rosten, R. Porter, and T. Drummond, "FASTER and Better: A Machine Learning Approach to Corner Detection," *IEEE TPAMI*, vol. 32, pp. 105–119, 2010.

[73] M. Agrawal, K. Konolige, and M. Blas, "CenSurE: Center Surround Extremas for Realtime Feature Detection and Matching," in *ECCV*, ser. Lecture Notes in Computer Science, 2008, vol. 5305, pp. 102–115.

[74] M. Ebrahimi and W. Mayol-Cuevas, "SUSurE: Speeded Up Surround Extrema Feature Detector and Descriptor for Realtime Applications," in *Workshop on Feature Detectors and Descriptors: "The State Of The Art and Beyond" as part of IEEE CVPR*, June 2009.

[75] "OpenCV Documentation," Accessed Jan 2017. [Online]. Available: http://docs.opencv.org/

[76] T. Akenine-Möller, E. Haines, and N. Hoffman, *Real-time Rendering*. CRC Press, 2008.

[77] J. Munkres, "Algorithms for the Assignment and Transportation Problems," *JSIAM*, vol. 5, no. 1, pp. 32–38, 1957.

[78] M. Brown, G. Hua, and S. Winder, "Discriminative Learning of Local Image Descriptors," *IEEE TPAMI*, vol. 33, no. 1, pp. 43–57, Jan 2011.

[79] K. Mikolajczyk and C. Schmid, "A Performance Evaluation of Local Descriptors," *IEEE TPAMI*, pp. 1615–1630, 2005.

[80] H. Hirschmüller and D. Scharstein, "Evaluation of Cost Functions for Stereo Matching," in *IEEE CVPR*, 2007, pp. 1–8.

[81] T. O. Aydın, N. Stefanoski, S. Croci, M. Gross *et al.*, "Temporally Coherent Local Tone Mapping of HDR Video," *ACM TOG*, vol. 33, no. 6, pp. 196:1–196:13, 2014.

[82] E. S. L. Gastal and M. M. Oliveira, "Domain Transform for Edge-Aware Image and Video Processing," *ACM TOG*, vol. 30, no. 4, pp. 69:1–69:12, 2011.

[83] F. Durand and J. Dorsey, "Fast Bilateral Filtering for the Display of High-dynamic-range Images," *ACM TOG*, pp. 257–266, July 2002.

[84] A. Criminisi, T. Sharp, C. Rother, and P. Pérez, "Geodesic Image and Video Editing," *ACM TOG*, vol. 29, no. 5, p. 134, 2010.

[85] C. Rhemann, A. Hosni, M. Bleyer, C. Rother *et al.*, "Fast Cost-Volume Filtering for Visual Correspondence and Beyond," in *IEEE CVPR*, 2011, pp. 3017–3024.

[86] M. Lang, O. Wang, T. Aydın, A. Smolic *et al.*, "Practical Temporal Consistency for Image-Based Graphics Applications," *ACM TOG*, vol. 31, no. 4, p. 34, 2012.

[87] G. Ye, E. Garces, Y. Liu, Q. Dai *et al.*, "Intrinsic Video and Applications," *ACM TOG*, vol. 33, no. 4, pp. 80:1–80:11, July 2014.

[88] N. Bonneel, J. Tompkin, K. Sunkavalli, D. Sun *et al.*, "Blind Video Temporal Consistency," *ACM TOG*, vol. 34, no. 6, 2015.

[89] C. Cigla and A. A. Alatan, "Information Permeability for Stereo Matching," *Signal Processing: Image Communication*, vol. 28, no. 9, pp. 1072–1088, 2013.

[90] E. Tola, V. Lepetit, and P. Fua, "A Fast Local Descriptor for Dense Matching," in *IEEE CVPR*, 2008, pp. 1–8.

[91] Y. Hu, R. Song, and Y. Li, "Efficient Coarse-to-Fine PatchMatch for Large Displacement Optical Flow," in *IEEE CVPR*, 2016.

[92] C. Barnes, E. Shechtman, A. Finkelstein, and D. Goldman, "PatchMatch: a Randomized Correspondence Algorithm for Structural Image Editing," *ACM TOG*, vol. 28, no. 3, p. 24, 2009.

[93] D. J. Butler, J. Wulff, G. B. Stanley, and M. J. Black, "A Naturalistic Open Source Movie for Optical Flow Evaluation," in *ECCV*. Springer, 2012, pp. 611–625.

[94] P. Milanfar, "A Tour of Modern Image Filtering: New Insights and Methods, Both Practical and Theoretical," *IEEE SPM*, vol. 30, no. 1, pp. 106–128, Jan 2013.

[95] C. Tomasi and R. Manduchi, "Bilateral Filtering for Gray and Color Images," in *ICCV*, Jan 1998, pp. 839–846.

[96] Z. Farbman, R. Fattal, D. Lischinski, and R. Szeliski, "Edge-preserving Decompositions for Multi-scale Tone and Detail Manipulation," *ACM TOG*, vol. 27, no. 3, pp. 67:1–67:10, Aug. 2008.

[97] R. Fattal, "Edge-Avoiding Wavelets and Their Applications," *ACM TOG*, vol. 28, no. 3, pp. 1–10, 2009.

[98] K. He, J. Sun, and X. Tang, "Guided Image Filtering," *IEEE TPAMI*, vol. 35, no. 6, pp. 1397–1409, June 2013.

[99] E. S. L. Gastal and M. M. Oliveira, "High-Order Recursive Filtering of Non-Uniformly Sampled Signals for Image and Video Processing," *CGF*, vol. 34, no. 2, pp. 81–93, May 2015, proceedings of Eurographics 2015.

[100] M. Aubry, S. Paris, S. W. Hasinoff, J. Kautz *et al.*, "Fast Local Laplacian Filters: Theory and Applications," *ACM TOG*, vol. 33, no. 5, p. 167, 2014.

[101] S. Paris, S. W. Hasinoff, and J. Kautz, "Local Laplacian Filters: Edge-Aware Image Processing With a Laplacian Pyramid," *ACM TOG*, vol. 30, no. 4, p. 68, 2011.

[102] P. Perona and J. Malik, "Scale-Space and Edge Detection Using Anisotropic Diffusion," *IEEE TPAMI*, vol. 12, no. 7, July 1990.

[103] D. Sun, S. Roth, and M. J. Black, "Secrets of Optical Flow Estimation and their Principles," in *IEEE CVPR*, June 2010, pp. 2432–2439.

[104] H. Zimmer, A. Bruhn, and J. Weickert, "Optic Flow in Harmony," *IJCV*, vol. 93, no. 3, pp. 368–388, 2011.

[105] R. Timofte and L. Van Gool, "Sparse Flow: Sparse Matching for Small to Large Displacement Optical Flow," in *IEEE WACV*, 2015.

[106] J. Revaud, P. Weinzaepfel, Z. Harchaoui, and C. Schmid, "EpicFlow: Edge-preserving Interpolation of Correspondences for Optical Flow," in *IEEE CVPR*, 2015, pp. 1164–1172.

[107] P. Weinzaepfel, J. Revaud, Z. Harchaoui, and C. Schmid, "Deepflow: Large Displacement Optical Flow with Deep Matching," in *IEEE ICCV*, 2013, pp. 1385–1392.

[108] J. Wulff and M. J. Black, "Efficient Sparse-to-Dense Optical Flow Estimation Using a Learned Basis and Layers," in *IEEE CVPR*, 2015, pp. 120–130.

[109] C. Liu, J. Yuen, and A. Torralba, "SIFT Flow: Dense Correspondence across Scenes and its Applications," *IEEE TPAMI*, 2010.

[110] W. Zhou, H. Li, M. Wang, Y. Lu *et al.*, "Binary Sift: Towards Efficient Feature Matching Verification for Image Search," in *ICIMCS*, 2012, pp. 1–6.

[111] N. Dalal and B. Triggs, "Histograms of Oriented Gradients for Human Detection," in *IEEE CVPR*, vol. 1, 2005, pp. 886–893.

[112] T. Trzcinski, M. Christoudias, and V. Lepetit, "Learning Image Descriptors with Boosting," *IEEE TPAMI*, vol. 37, no. 3, pp. 597–610, 2015.

[113] L. Baroffio, M. Cesana, A. Redondi, and M. Tagliasacchi, "Bamboo: A Fast Descriptor Based on AsymMetric Pairwise BOOsting," in *IEEE ICIP*, 2014, pp. 5686–5690.

[114] X. Yang and K. T. Cheng, "Learning Optimized Local Difference Binaries for Scalable Augmented Reality on Mobile Devices," *IEEE TVCG*, vol. 20, no. 6, pp. 852–865, June 2014.

[115] T. Brox and J. Malik, "Large displacement optical flow: Descriptor matching in variational motion estimation," *IEEE PAMI*, vol. 33, no. 3, pp. 500–513, 2011.

[116] C. Strecha, A. Bronstein, M. Bronstein, and P. Fua, "LDAHash: Improved Matching with Smaller Descriptors," *IEEE TPAMI*, vol. 34, no. 1, pp. 66–78, Jan 2012.

[117] Y. Yamauchi and H. Fujiyoshi, "Binary code-based Human Detection," 2012.

[118] B. Jun, I. Choi, and D. Kim, "Local transform features and hybridization for accurate face and human detection," *IEEE TPAMI*, vol. 35, no. 6, pp. 1423–1436, 2013.

[119] H. Fu, H. Zhao, X. Kong, and X. Zhang, "BHoG: Binary Descriptor for Sketch-Based Image Retrieval," *Multimedia Systems*, vol. 22, no. 1, pp. 127–136, 2016.

[120] M. Zwicker, H. Pfister, J. V. Baar, and M. Gross, "EWA Splatting," *IEEE TVCG*, vol. 8, no. 3, pp. 223–238, 2002.

[121] K. Zuiderveld, "Contrast Limited Adaptive Histogram Equalization," in *Graphics Gems IV*, P. S. Heckbert, Ed. San Diego, CA, USA: Academic Press Professional, Inc., 1994, pp. 474–485.

[122] D. Nehab, A. Maximo, R. S. Lima, and H. Hoppe, "Gpu-efficient recursive filtering and summed-area tables," *ACM TOG*, vol. 30, no. 6, p. 176, 2011.

[123] P. Yu, X. Yang, and L. Chen, "Parallel-friendly Patch Match Based on Jump Flooding," in *Advances on Digital Television and Wireless Multimedia Communications*.   Springer, 2012, pp. 15–21.

[124] A. Vedaldi and B. Fulkerson, "VLFeat: An Open and Portable Library of Computer Vision Algorithms," Accessed Jan 2017. [Online]. Available: http://www.vlfeat.org

[125] M. Calonder, V. Lepetit, C. Strecha, and P. Fua, "BRIEF: Binary Robust Independent Elementary Features," in *ECCV*, ser. Lecture Notes in Computer Science, 2010, vol. 6314, pp. 778–792.

[126] K. He and J. Sun, "Computing Nearest-Neighbor Fields via Propagation-Assisted kd-Trees," in *IEEE CVPR*.   IEEE, 2012, pp. 111–118.

[127] C. Bailer, B. Taetz, and D. Stricker, "Flow Fields: Dense Correspondence Fields for Highly Accurate Large Displacement Optical Flow Estimation," in *IEEE ICCV*, 2015, pp. 4015–4023.

[128] R. Kennedy and C. J. Taylor, "Optical Flow with Geometric Occlusion Estimation and Fusion of Multiple Frames," in *EMMCVPR*, 2015.

[129] M. Menze, C. Heipke, and A. Geiger, "Discrete Optimization for Optical Flow," in *GCPR*.   Springer, 2015, pp. 16–28.

[130] P. Greisen, M. Lang, S. Heinzle, and A. Smolic, "Algorithm and VLSI Architecture for Real-Time 1080P60 Video Retargeting," in *ACM EGGH-HPG*, 2012, pp. 57–66.

[131] D. Scharstein, H. Hirschmüller, Y. Kitajima, G. Krathwohl *et al.*, "High-Resolution Stereo Datasets with Subpixel-Accurate Ground Truth," in *GCPR*, 2014, pp. 31–42.

[132] C. Kerl, J. Sturm, and D. Cremers, "Dense Visual SLAM for RGB-D Cameras," in *IEEE IROS*, Nov 2013.

[133] D. Krishnan and R. Szeliski, "Multigrid and Multilevel Preconditioners for Computational Photography," in *ACM TOG*, vol. 30, no. 6, 2011.

[134] I. Koutis, G. L. Miller, and D. Tolliver, "Combinatorial Preconditioners and Multilevel Solvers for Problems in Computer Vision and Image Processing," *JCVIU*, vol. 115, no. 12, pp. 1638–1646, 2011.

[135] A. Björck, "Numerical Methods for Least Squares Problems," *SIAM*, 1996.

[136] Y. Saad, *Iterative Methods for Sparse Linear Systems Second Edition*. Philadelphia, USA: SIAM, 2003.

[137] R. Barrett, M. Berry, T. Chan, J. Demmel *et al.*, *Templates for the Solution of Linear Systems: Building Blocks for Iterative Methods*. Philadelphia, USA: SIAM, 1994.

[138] J. W. Demmel, *Applied Numerical Linear Algebra*. Philadelphia, USA: SIAM, 1997.

[139] A. Roldao and G. A. Constantinides, "A High Throughput FPGA-Based Floating Point Conjugate Gradient Implementation for Dense Matrices," *ACM TRETS*, vol. 3, no. 1, Jan 2010.

[140] O. Maslennikow, V. Lepekha, A. Sergiyenko, A. Tomas *et al.*, "Parallel Implementation of Cholesky LL^T-Algorithm in FPGA-based Processor," in *ICPPAM*. Springer, 2007, pp. 137–147.

[141] H. Cho, J. Lee, and Y. Kim, "Efficient Implementation of Linear System Solution Block Using LDL^T Factorization," *SoC 2008*, vol. 03, 2008.

[142] J. Sun, G. Peterson, and O. Storaasli, "High-performance Mixed-Precision Linear Solver for FPGAs," *IEEE TC*, vol. 57, no. 12, 2008.

[143] D. Yang, G. D. Peterson, and H. Li, "Compressed Sensing and Cholesky Decomposition on FPGAs and GPUs," *Parallel Computing*, vol. 38, no. 8, pp. 421–437, 2012.

[144] U. Vishnoi and T. Noll, "Cross-Layer Optimization of QRD Accelerators," in *ESSCIRC*, 2013.

[145] Micron, "LPDDR2 System-Power Calculator," Accessed Jan 2017. [Online]. Available: http: //www.micron.com/support/tools-and-utilities/power-calc

[146] S. Thoziyoor, N. Muralimanohar, J. H. Ahn, and N. P. Jouppi, "Cacti 5.1," Technical Report HPL-2008-20, HP Labs, Tech. Rep., 2008.

[147] S. Li, J. H. Ahn, R. D. Strong, J. B. Brockman *et al.*, "McPAT: An Integrated Power, Area, and Timing Modeling Framework for Multicore and Manycore Architectures," in *IEEE/ACM MICRO*, 2009, pp. 469–480.

[148] T. Yang and H. X. Lin, "Solving Sparse Least Squares Problems with Preconditioned CGLS Method on Parallel Distributed Memory Computers," *Parallel Algorithms and Applications*, vol. 13, no. 4, pp. 289–305, 1999.

[149] C. C. Paige and M. A. Saunders, "LSQR: An Algorithm for Sparse Linear Equations and Sparse Least Squares," *ACM transactions on mathematical software*, vol. 8, no. 1, pp. 43–71, 1982.

[150] P. Greisen, M. Runo, P. Guillet, S. Heinzle *et al.*, "Evaluation and FPGA Implementation of Sparse Linear Solvers for Video Processing Applications," *IEEE TCSVT*, vol. 23, no. 8, pp. 1402–1407, 2013.

[151] A. George, "Nested Dissection of a Regular Finite Element Mesh," *SINUM*, 1973.

[152] P. Amestoy, T. Davis, and I. Duff, "Algorithm 837: AMD, An Approximate Minimum Degree Ordering Algorithm," *ACM TOMS*, vol. 30, no. 3, 2004.

[153] B. Boroujerdian, B. Keller, and Y. Lee, "LPDDR2 Memory Controller Design in a 28 nm Process," 2012, Accessed Jan 2017. [Online]. Available: http://www.eecs.berkeley.edu/~bkeller/rekall.pdf

[154] F. Ross, "A Review of LPDDR3 Commands, Operations & Functions," 2012. [Online]. Available: http://www.jedec.org/lpddr3-presentations

[155] RMIT University, "An Uncompressed Stereoscopic 3D HD Video Library," Accessed Jan 2017. [Online]. Available: http://www.rmit3dv.com

[156] J. Dongarra, I. Duff, D. Sorensen, and H. van der Vorst, *Numerical Linear Algebra for High-Performance Computers*. Philadelphia, USA: SIAM, 1998.

[157] R. J. Baker, "What Memory Wall?" *IEEE Solid-State Circuits Magazine*, vol. 8, no. 2, Spring 2016.

[158] Xilinx, "UltraRAM: Breakthrough Embedded Memory Integration on UltraScale+ Devices," Tech. Rep., June 2016, wP477 (v1.0). [Online]. Available: https://www.xilinx.com/support/documentation/white_papers/wp477-ultraram.pdf

[159] S.-C. T. Choi, C. C. Paige, and M. A. Saunders, "MINRES-QLP: A Krylov Subspace Method for Indefinite or Singular Symmetric Systems," *SIAM Journal on Scientific Computing*, vol. 33, no. 4, pp. 1810–1836, 2011.

[160] D. C.-L. Fong and M. A. Saunders, "CG Versus MINRES: An Empirical Comparison," *SQU Journal for Science*, vol. 17, no. 1, pp. 44–62, 2012.

[161] W. Briggs, V. Henson, and S. McCormick, *A Multigrid Tutorial*, 2nd ed. SIAM, 2000.

[162] K. Palem and A. Lingamneni, "Ten Years of Building Broken Chips: The Physics and Engineering of Inexact Computing," *ACM TECS*, 2013.

[163] J. Konrad and P. Agniel, "Artifact Reduction in Lenticular Multiscopic 3D Displays by Means of Anti-Alias Filtering," *Stereoscopic Displays and Virtual Reality Systems*, vol. 5006, no. 1, pp. 336–347, 2003.

[164] J. Konrad and P. Agniel, "Non-Orthogonal Subsampling and Anti-Alias Filtering for Multiscopic 3D Displays," *Stereoscopic Displays and Virtual Reality Systems XI*, vol. 5291, no. 1, pp. 105–116, 2004.

[165] J. Konrad and P. Agniel, "Subsampling Models and Anti-Alias Filters for 3-D Automultiscopic Displays," *IEEE TIP*, vol. 15, no. 1, pp. 128–140, Jan 2006.

[166] J. Konrad and A. Jain, "Crosstalk in Automultiscopic 3-D Displays: Blessing in Disguise?" *Stereoscopic Displays and Virtual Reality Systems XIV*, vol. 6490, no. 1, p. 649012, 2007.

[167] P. Heckbert, "Fundamentals of Texture Mapping and Image Warping," Masters Thesis, Univ. of California, Berkeley, Dept. of Electrical Eng. and Computer Science, 1989.

[168] N. A. Dodgson, "Technical Report: Image Resampling," University of Cambridge, 15 JJ Thomson Avenue, Cambridge, UK, Tech. Rep., Aug 1992.

[169] Alioscopy, "Alioscopy 3D Displays," Accessed Jan 2017. [Online]. Available: http://www.alioscopy.com/en/3Ddisplays.php

[170] IEC, "Call for Proposals on 3D Video Coding Technology," ISO/IEC JTC1/SC29/WG11, Geneva, Switzerland, Tech. Rep. MPEG2011/N12036, March 2011, approved.

[171] M. U. Rick Szeliski, Simon Winder, "High-Quality Multi-Pass Image Resampling," Tech. Rep., Feb 2010. [Online]. Available: https://www.microsoft.com/en-us/research/publication/high-quality-multi-pass-image-resampling

[172] R.-P. M. Berretty, F. J. Peters, and G. T. G. Volleberg, "Real-Time Rendering for Multiview Autostereoscopic Displays," *Proc. SPIE 6055, Stereoscopic Displays and Virtual Reality Systems XIII*, 2006.

[173] F. M. Khan, M. G. Arnold, and W. M. Pottenger, "Hardware-Based Support Vector Machine Classification in Logarithmic Number Systems," in *IEEE ISCAS*, 2005, pp. 5154–5157 Vol. 5.

[174] S. Afifi, H. Gholam Hosseini, and R. Sinha, *PSIVT 2015 Workshops. Revised Selected Papers.* Springer International Publishing, 2016, ch. Hardware Acceleration of SVM-Based Classifier for Melanoma Images, pp. 235–245.

[175] M. Komorkiewicz, M. Kluczewski, and M. Gorgon, "Floating Point HOG Implementation for Real-Time Multiple Object Detection," in *FPL*, 2012, pp. 711–714.

[176] I. Hong, G. Kim, Y. Kim, D. Kim *et al.*, "A 27 mW Reconfigurable Marker-Less Logarithmic Camera Pose Estimation Engine for Mobile Augmented Reality Processor," *IEEE JSSC*, vol. 50, no. 11, pp. 2513–2523, Nov 2015.

[177] G. Frantz and R. Simar, "Comparing Fixed- and Floating-Point DSPs," 2004, Texas Instruments, Dallas, TX, USA, Accessed Jan 2017. [Online]. Available: http://www.ti.com.cn/cn/lit/wp/spry061/spry061.pdf

[178] Altera, "The Industry's First Floating-Point FPGA," Accessed Jan 2017. [Online]. Available: http://www.altera.com/en_US/pdfs/literature/po/bg-floating-point-fpga.pdf

[179] ARM Ltd., Cambridge, UK, "Cortex-M4 Processor," Accessed Jan 2017. [Online]. Available: http://www.arm.com/products/processors/cortex-m/cortex-m4-processor.php

[180] N. Kingsbury and P. Rayner, "Digital Filtering Using Logarithmic Arithmetic," *Electronics Letters*, vol. 7, no. 2, pp. 56–58, Jan 1971.

[181] E. Swartzlander and A. Alexopoulos, "The Sign/Logarithm Number System," *IEEE TOC*, vol. C-24, no. 12, pp. 1238–1242, Dec 1975.

[182] V. Paliouras and T. Stouraitis, "A Novel Algorithm for Accurate Logarithmic Number System Subtraction," in *IEEE ISCAS*, 1996, pp. 268–271.

[183] M. G. Arnold, T. A. Bailey, J. R. Cowles, and M. D. Winkel, "Arithmetic Co-transformations in the Real and Aomplex Logarithmic Number Systems," *IEEE TOC*, vol. 47, no. 7, pp. 777–786, 1998.

[184] J. Coleman, "Simplification of Table Structure in Logarithmic Arithmetic," *Electronics Letters*, vol. 31, no. 22, pp. 1905–1906, Oct 1995.

[185] J. N. Coleman, E. Chester, C. I. Softley, and J. Kadlec, "Arithmetic on the European Logarithmic Microprocessor," *IEEE TOC*, vol. 49, no. 7, pp. 702–715, 2000.

[186] P. Vouzis, S. Collange, and M. Arnold, "LNS Subtraction Using Novel Cotransformation and/or Interpolation," in *IEEE ASAP*, 2007, pp. 107–114.

[187] J. N. Coleman, C. I. Softley, J. Kadlec, R. Matousek *et al.*, "The European Logarithmic Microprocessor," *IEEE TOC*, vol. 57, no. 4, pp. 532–546, 2008.

[188] R. Ismail and J. Coleman, "ROM-less LNS," in *IEEE ARITH*, July 2011.

[189] J. N. Coleman and R. C. Ismail, "LNS with Co-Transformation Competes with Floating-Point," *IEEE TOC*, vol. 65, no. 1, pp. 136–146, Jan 2016.

[190] H. Esmaeilzadeh, A. Sampson, L. Ceze, and D. Burger, "Architecture Support for Disciplined Approximate Programming," in *ACM SIGPLAN*, vol. 47, no. 4, 2012, pp. 301–312.

[191] S. Venkataramani, V. K. Chippa, S. T. Chakradhar, K. Roy *et al.*, "Quality Programmable Vector Processors for Approximate Computing," in *IEEE/ACM MICRO*, 2013, pp. 1–12.

[192] E. Lindholm, J. Nickolls, S. Oberman, and J. Montrym, "NVIDIA Tesla: A Unified Graphics and Computing Architecture," *IEEE MICRO*, vol. 28, no. 2, pp. 39–55, March 2008.

[193] A. Ukil, V. H. Shah, and B. Deck, "Fast Computation of Arctangent Functions for Embedded Applications: A Comparative Analysis," in *IEEE ISIE*, June 2011, pp. 1206–1211.

[194] F. de Dinechin and A. Tisserand, "Multipartite Table Methods," *IEEE TOC*, vol. 54, no. 3, pp. 319–330, March 2005.

[195] J. Detrey and F. de Dinechin, "Table-based Polynomials for Fast Hardware Function Evaluation," in *IEEE ASAP*, July 2005, pp. 328–333.

[196] J. Detrey and F. de Dinechin, "A Tool for Unbiased Comparison Between Logarithmic and Floating-Point Arithmetic," *JSPS*, vol. 49, no. 1, pp. 161–175, 2007.

[197] J. Rust, F. Ludwig, and S. Paul, "Low Complexity QR-Decomposition Architecture using the Logarithmic Number System," in *IEEE DATE*, 2013, pp. 97–102.

[198] J. Garcia, M. G. Arnold, L. Bleris, and M. V. Kothare, "LNS Architectures for Embedded Model Predictive Control Processors," in *ACM CASES*, 2004, pp. 79–84.

[199] D. Miyashita, E. H. Lee, and B. Murmann, "Convolutional Neural Networks using Logarithmic Data Representation," *ArXiv*, March 2016.

[200] I. Kouretas, C. Basetas, and V. Paliouras, "Low-Power Logarithmic Number System Addition/Subtraction and Their Impact on Digital Filters," *IEEE TOC*, vol. 62, no. 11, pp. 2196–2209, 2013.

[201] M. Arnold and S. Collange, "A Real/Complex Logarithmic Number System ALU," *IEEE TOC*, vol. 60, no. 2, pp. 202–213, Feb 2011.

[202] M. G. Arnold, J. Cowles, V. Paliouras, and I. Kouretas, "Towards a Quaternion Complex Logarithmic Number System," in *IEEE ARITH*, 2011, pp. 33–42.

[203] R. Ismail, M. Zakaria, and S. Murad, "Hybrid Logarithmic Number System Arithmetic Unit: A Review," in *IEEE ICCAS*, 2013, pp. 55–58.

[204] H. Fu, O. Mencer, and W. Luk, "Optimizing Logarithmic Arithmetic on FPGAs," in *IEEE FCCM*, 2007, pp. 163–172.

[205] H. Fu, O. Mencer, and W. Luk, "FPGA Designs With Optimized Logarithmic Arithmetic," *IEEE TOC*, vol. 7, no. 59, pp. 1000–1006, 2010.

[206] F. de Dinechin and B. Pasca, "Designing Custom Arithmetic Data Paths with FloPoCo," *IEEE Design & Test of Computers*, vol. 28, no. 4, pp. 18–27, 2011, Accessed Jan 2017. [Online]. Available: http://flopoco.gforge.inria.fr

[207] J.-M. Muller, *Elementary Functions*. Springer, 2006.

[208] T.-J. Kwon, J. Sondeen, and J. Draper, "Design Trade-Offs in Floating-Point Unit Implementation for Embedded and

Processing-in-Memory Systems," in *IEEE ISCAS*, 2005, pp. 3331–3334.

[209] K. Karuri, R. Leupers, G. Ascheid, H. Meyr *et al.*, "Design and Implementation of a Modular and Portable IEEE 754 Compliant Floating-Point Unit," in *IEEE DATE*, vol. 2, 2006, pp. 1–6.

[210] S. Galal and M. Horowitz, "Energy-Efficient Floating-Point Unit Design," *IEEE TOC*, vol. 60, no. 7, pp. 913–922, July 2011.

[211] S. Galal, O. Shacham, J. S. B. II, J. Pu *et al.*, "FPU Generator for Design Space Exploration," in *IEEE ARITH*, April 2013, pp. 25–34.

[212] S. F. Oberman and M. Y. Siu, "A High-Performance Area-Efficient Multifunction Interpolator," in *IEEE ARITH*, June 2005, pp. 272–279.

[213] D. D. Caro, N. Petra, and A. G. M. Strollo, "High-Performance Special Function Unit for Programmable 3-D Graphics Processors," *IEEE TCAS-I*, vol. 56, no. 9, pp. 1968–1978, Sept 2009.

[214] B. G. Nam, H. Kim, and H. J. Yoo, "A Low-Power Unified Arithmetic Unit for Programmable Handheld 3-D Graphics Systems," *IEEE JSSC*, vol. 42, no. 8, pp. 1767–1778, Aug 2007.

[215] F. de Dinechin, M. Istoan, and G. Sergent, "Fixed-point Trigonometric Functions on FPGAs," *SIGARCH Comput. Archit. News*, vol. 41, no. 5, pp. 83–88, June 2014.

[216] F. de Dinechin and M. Istoan, "Hardware Implementations of Fixed-Point Atan2," in *IEEE ARITH*, June 2015, pp. 34–41.

[217] J. F. Hart, *Computer Approximations*. Krieger Publishing Co., 1978.

[218] D. Rossi, A. Pullini, I. Loi, M. Gautschi *et al.*, "A 60 GOPS/W, -1.8 V to 0.9 V Body Bias ULP Cluster in 28nm UTBB FD-SOI Technology ," *Solid-State Electronics*, vol. 117, pp. 170 – 184, 2016.

[219] R. Gutierrez, V. Torres, and J. Valls, "FPGA-Implementation of Atan(Y/X) Based on Logarithmic Transformation and

LUT-based Techniques," *JSA*, vol. 56, no. 11, pp. 588–596, Nov. 2010.

[220] P. K. Meher, J. Valls, T. B. Juang, K. Sridharan *et al.*, "50 Years of CORDIC: Algorithms, Architectures, and Applications," *IEEE TCAS-I*, vol. 56, no. 9, pp. 1893–1907, Sept 2009.

[221] S. Chevillard, M. Joldeş, and C. Lauter, "Sollya: An Environment for the Development of Numerical Codes," in *ICMS*. Springer, 2010, pp. 28–31.

[222] F. De Dinechin, M. Joldes, and B. Pasca, "Automatic Generation of Polynomial-Based Hardware Architectures for Function Evaluation," in *IEEE ASAP*, 2010, pp. 216–222.

[223] M. Gautschi, A. Traber, A. Pullini, L. Benini *et al.*, "Tailoring Instruction-Set Extensions for an Ultra-Low Power Tightly-Coupled Cluster of OpenRISC Cores," in *IFIP/IEEE VLSI-SoC*, 2015, pp. 25–30.

[224] H. Kaul, M. Anders, S. Mathew, S. Hsu *et al.*, "A 1.45GHz 52-to-162GFLOPS/W Variable-Precision Floating-Point Fused Multiply-Add Unit With Certainty Tracking in 32nm CMOS," in *IEEE ISSCC*, Feb 2012, pp. 182–184.

[225] Wikipedia, "Fast Inverse Square Root," Accessed Jan 2017. [Online]. Available: http://en.wikipedia.org/wiki/Fast_inverse_square_root

[226] J.-M. Muller, N. Brisebarre, F. De Dinechin, C.-P. Jeannerod *et al.*, *Handbook of Floating-Point Arithmetic*. Springer Science & Business Media, 2009.

[227] D. Monniaux, "The Pitfalls of Verifying Floating-point Computations," *ACM TOPLAS*, vol. 30, no. 3, pp. 12:1–12:41, May 2008.

[228] K. Müller, A. Vetro, and V. Baroncini, "Report of subjective test results from the call for proposals on 3d video coding," ISO/IEC JTC1/SC29/WG11, Geneva, Switzerland, Tech. Rep. MPEG2011/N12347, Nov 2011.

[229] N. Stefanoski, M. Lang, and A. Smolic, "Image Quality vs. Rate Optimized Coding of Warps for View Synthesis in 3D Video Applications," in *IEEE ICIP*, Sept 2012, pp. 1289–1292.

[230] H. Kim, S. Lee, and A. Bovik, "Saliency Prediction on Stereoscopic Videos," *IEEE TIP*, vol. 23, no. 4, pp. 1476–1490, April 2014.

[231] M. Zwicker, W. Matusik, F. Durand, and H. Pfister, "Antialiasing for Automultiscopic 3D Displays," in *ACM EGSR*, 2006, pp. 73–82.

[232] A. Chapiro, S. Heinzle, T. O. Aydın, S. Poulakos *et al.*, "Optimizing Stereo-to-Multiview Conversion for Autostereoscopic Displays," *CGF*, vol. 33, no. 2, 2014.

[233] C. Riechert, F. Zilly, P. Kauff, J. Güther *et al.*, "Fully Automatic Stereo-to-Multiview Conversion in Autostereoscopic Displays," *The best of IET and IBC*, vol. 4, no. 8, p. 14, 2012.

[234] C. Liao, H. Yeh, K. Zhang, V. Geert *et al.*, "Stereo Matching and Viewpoint Synthesis FPGA Implementation," in *3D-TV System with Depth-Image-Based Rendering*. Springer New York, 2013.

[235] J. Perez, P. Sanchez, and M. Martinez, "High Memory Throughput FPGA Architecture for High-Definition Belief-Propagation Stereo Matching," in *SCS*, Nov 2009, pp. 1–6.

[236] N.-C. Chang, T.-H. Tsai, B.-H. Hsu, Y.-C. Chen *et al.*, "Algorithm and Architecture of Disparity Estimation With Mini-Census Adaptive Support Weight," in *IEEE TCSVT*, vol. 20, no. 6, June 2010, pp. 792–805.

[237] L. Zhang, K. Zhang, T. S. Chang, G. Lafruit *et al.*, "Real-time High-definition Stereo Matching on FPGA," in *ACM/SIGDA FPGA*. ACM, 2011, pp. 55–64.

[238] A. Akın, I. Baz, A. Schmid, and Y. Leblebici, "Dynamically Adaptive Real-Time Disparity Estimation Hardware using Iterative Refinement," *Integration*, vol. 47, no. 3, pp. 365–376, 2014.

[239] M. Werner, B. Stabernack, and C. Riechert, "Hardware Implementation of a Full HD Real-Time Disparity Estimation Algorithm," *IEEE TCE*, vol. 60, no. 1, pp. 66–73, Feb 2014.

[240] H.-J. Chen, F.-H. Lo, F. Jan, and S.-D. Wu, "Real-time Multi-View Rendering Architecture for Autostereoscopic Displays," in *IEEE ISCAS*, May 2010, pp. 1165–1168.

[241] B. Fan, Q. Kong, X. Yuan, Z. Wang *et al.*, "Learning Weighted Hamming Distance for Binary Descriptors," in *IEEE ICASSP*, May 2013, pp. 2395–2399.

[242] Y. Horng, Y. Tseng, and T. Chang, "VLSI Architecture for Real-Time HD1080p View Synthesis Engine," *IEEE TCSVT*, vol. 21, no. 9, 2011.

[243] P.-K. Tsung, P.-C. Lin, K.-Y. Chen, T.-D. Chuang *et al.*, "A 216fps 4096x2160p 3DTV Set-Top Box SoC for Free-Viewpoint 3DTV Applications," in *IEEE ISSCC*, 2011, pp. 124–126.

[244] F.-J. Chang, Y.-C. Tseng, and T.-S. Chang, "A 94fps View Synthesis Engine for HD1080p Video," in *IEEE VCIP*, 2011.

[245] F. De Dinechin, B. Pasca, O. Cret, and R. Tudoran, "An FPGA-Specific Approach to Floating-point Accumulation and Sum-of-products," in *IEEE ICECE*, 2008, pp. 33–40.

[246] S. Demirsoy and M. Langhammer, "Cholesky Decomposition Using Fused Datapath Synthesis," in *ACM/SIGDA FPGA*, 2009.

[247] A. L. Shimpi, "The iPad Air Review," Oct 2013, Accessed Jan 2017. [Online]. Available: http://www.anandtech.com/show/7460/apple-ipad-air-review

[248] D. A. Patterson and J. L. Hennessy, *Computer Organization and Design: The Hardware/Software Interface.* Newnes, 2013.

[249] R. Nair, "Evolution of Memory Architecture," *PIEEE*, vol. 103, no. 8, pp. 1331–1345, Aug 2015.

[250] M. B. Taylor, "Is Dark Silicon Useful?: Harnessing the Four Horsemen of the Coming Dark Silicon Apocalypse," in *DAC*. ACM, 2012, pp. 1131–1136.

[251] E. Azarkhish, C. Pfister, D. Rossi, I. Loi *et al.*, "Logic-Base Interconnect Design for Near Memory Computing in the Smart Memory Cube," *IEEE TVLSI*, vol. 25, no. 1, pp. 210–223, Jan 2017.

[252] Kaufmann, Peter and Wang, Oliver and Sorkine-Hornung, Alexander and Sorkine-Hornung, Olga and Smolic, Aljoscha and Gross, Markus, "Finite element image warping," in *Computer Graphics Forum*, vol. 32, no. 2pt1, 2013, pp. 31–39.

[253] S.-J. Yao, L.-H. Wang, C.-L. Lin, and M. Zhang, "Real-time Stereo to Multi-view Conversion System Based on Adaptive Meshing," *Journal of Real-Time Image Processing*, pp. 1–19, 2015.

[254] V. K. Adhikarla, P. T. Kovács, A. Barsi, T. Balogh *et al.*, "View Synthesis for Lightfield Displays Using Region Based Non-linear Image Warping," in *IC3D*, Dec 2012, pp. 1–6.

[255] P. Klavík, A. C. I. Malossi, C. Bekas, and A. Curioni, "Changing Computing Paradigms Towards Power Efficiency," *Phil Trans.: Mathematical, Physical and Engineering Sciences*, vol. 372, no. 2018, 2014.

[256] J. Bates, "Processing With Compact Arithmetic Processing Element," April 2012, uS Patent 8,150,902.

[257] R. Eaton, J. McBride, and J. Bates, "Reliable ISR Algorithms for a Very-low-power Approximate Computer," in *SPIE DSS*, 2013.

[258] A. Sampson, W. Dietl, E. Fortuna, D. Gnanapragasam *et al.*, "EnerJ: Approximate Data Types for Safe and General Low-Power Computation," *SIGPLAN Not.*, vol. 46, no. 6, June 2011.

[259] J. Kim, M. Sullivan, E. Choukse, and M. Erez, "Bit-Plane Compression: Transforming Data for Better Compression in Many-Core Architectures," in *ACM/IEEE ISCA*, June 2016, pp. 329–340.

[260] Xilinx, "All Programmable Heterogeneous MPSoC," Accessed Jan 2017. [Online]. Available: http://www.xilinx.com/products/silicon-devices/soc/zynq-ultrascale-mpsoc

Curriculum Vitae

Michael Schaffner was born in Zurich, Switzerland, in 1986. He received his BSc and MSc degrees from the Swiss Federal Institute of Technology Zurich, Switzerland, in 2009 and 2012. Since late 2012, he has been a research assistant with the Integrated Systems Laboratory and with Disney Research Zurich. His research interests include digital signal processing, video processing, and the design of very large scale integration circuits and systems. Michael Schaffner received the ETH Medal for his diploma thesis in 2013.

SERIES IN MICROELECTRONICS

Edited by Qiuting Huang, Mathieu Maurice Luisier,
Andreas Schenk, Bernd Witzigmann

More volumes of this series at www.hartung-gorre.de

www.ingramcontent.com/pod-product-compliance
Lightning Source LLC
Chambersburg PA
CBHW060342220326
41598CB00023B/2782

9783866286245